George Edgar Ladd

A Preliminary Report on a Part of the Clays of Georgia

Bulletin No. 6-A

George Edgar Ladd

A Preliminary Report on a Part of the Clays of Georgia
Bulletin No. 6-A

ISBN/EAN: 9783337059392

Printed in Europe, USA, Canada, Australia, Japan

Cover: Foto ©berggeist007 / pixelio.de

More available books at **www.hansebooks.com**

GEOLOGICAL SURVEY OF GEORGIA

W. S. YEATES, State Geologist

BULLETIN NO. 6—A

A PRELIMINARY REPORT

ON A PART OF THE

CLAYS OF GEORGIA

GEO. E. LADD

Assistant Geologist

1898

THE ADVISORY BOARD

of the Geological Survey of Georgia

(Ex-Officio)

His Excellency, W. Y. ATKINSON, Governor of Georgia,

President of the Board

Hon. R. T. NESBITT----------Commissioner of Agriculture

Hon. G. R. GLENN----------Commissioner of Public Schools

Hon. W. J. SPEER------------------------State Treasurer

Hon. W. A. WRIGHT--------------------Comptroller-General

Hon. WILLIAM CLIFTON---------------Secretary of State

Hon. J. M. TERRELL--------------------Attorney-General

October 4th, 1898.

To His Excellency, W. Y. ATKINSON, *Governor, and President of the Advisory Board of the Geological Survey of Georgia:* —

SIR: — I have the honor, to transmit, herewith, the report of Dr. Geo. E. Ladd, formerly an Assistant Geologist on this Survey, on a part of THE CLAYS OF GEORGIA. As stated in my Administrative Report for 1897, the first part of the report, which had been furnished the State Printer, was withdrawn by Dr. Ladd, in March, 1897, to add new matter. He resigned his position the following October, without finishing the report; and I am just in receipt of the last pages, which complete it.

The subject is one of very great economic importance to the State; and I trust the report will meet, in part, at least, the many requests for information about the clays of Georgia, which I have received from investors, in the North and West. Field-work will be continued on this subject, at the earliest time possible; and the second bulletin of the series, which shall cover more territory, will be published.

Very respectfully yours,
W. S. YEATES,
State Geologist.

THE CLAYS OF GEORGIA

CHAPTER I

GENERAL REMARKS ON CLAYS

Clays and clay industries may be conveniently discussed, from scientific, industrial or historical standpoints.

Scientific considerations deal with the definition and classification of clays; their geological occurrence and distribution; their geographical distribution; their chemical and physical composition and structure; their origin and properties; and the composition, structure and properties of clay-products.

Industrial considerations are those, dealing, specifically, with the manufacture of clay-wares, for all the various uses, such as structures (buildings, roads etc.), ornaments (architectural pieces and art-pottery), fire-proofing, heat and electricity insulating, utensils (pottery and porcelain), sewer-pipes, drain-tiles and miscellaneous products; also, with the use, or misuse, of clay, as an adulterant (for paints, foods etc.). They, further, deal with the methods of mining and preparing clays, and with the statistics, connected with all these subjects.

Historical considerations are those, dealing with the history of the manufacture of clay-products, and with the development of our knowledge of clays.

Clay articles are among the earliest manufactured products of primitive man; and fragments of clay utensils are among the most enduring records of man's existence. The most ancient historical records refer to the art of pottery-making, as being then in an advanced stage. The Chinese claim, that glazed pottery was invented, by an Emperor, called Hwang-to, whose "legendary reign, of a hundred years, is said to have commenced in the year 2,697, B.C."

It is claimed, that the potter's-wheel was known to the Egyptians, as early as 2,500 years, B. C.; and that a sort of porcelain was manufactured by them, during the Sixth Egyptian Dynasty, which is stated, by some authors, to have been as early as 3,700, B. C. From these early beginnings, the use of clay has gradually expanded, increasing rapidly, since the discovery of the art of making porcelain, by the Chinese, at about the beginning of the Christian era; and, again, enormously, on the introduction of porcelain into Europe; and still more emphatically, since the art of making white porcelain was discovered by the Europeans, at the beginning of the Eighteenth century.

At the present time, the uses, to which clay is put, have become so varied, and its application to many of these uses, so universal, that this has been called, not unjustly, the "Age of Clay", as well as the "Age of Steel".

In spite of the varied application of clays to the technical and fine arts, ranging all the way, in result, from the coarsest common brick to the most exquisite pieces of artistic pottery, there has been an extraordinarily small amount of scientific research, concerning their nature and properties. What has been done, has been achieved largely by the Germans; though, at present, the importance of the subject is becoming recognized; and the Americans, English, French, Swedes and Russians are beginning to turn their attention to these scientific investigations. In this country, recognition of the subject is illustrated, by the number of States, which have finished, or are at present undertaking, the preparation of treatises on the occurrence, distribution, origin, nature, uses etc. of the clays, within their boundaries. Among these, are the States of New York, New Jersey, Maryland, Ohio, Kentucky, Iowa, Missouri, Indiana, California, Arkansas, Texas, Alabama and Georgia. The reports of some are already published. Those of others are in print, and the rest are unfinished, awaiting the completion of field-work.

DEFINITION AND CLASSIFICATION

The term clay is a generic name, for a group of more or less consolidated secondary rocks, which are composed of hydrous silicates of aluminum, usually associated with other minerals (present as impurities), which originate, chemically, as products, after the decomposition of various aluminum-bearing minerals, chiefly feldspar, and which occur, either at the place of their origin, from the maternal rock, or removed thence by geologic agencies, in foreign localities.

Clays are commonly classified, from a commercial standpoint, with reference, especially, to the uses, for which particular clays are adaptable. From a geological standpoint, the writer suggests, that they be classified into larger divisions, based on origin and mode of occurrence; and into minor subdivisions, according to their composition and properties. The larger divisions of such a classification would be somewhat as follows: —

CLASSIFICATION OF CLAYS

INDIGENOUS. —
    A. Kaolins.
        a. Superficial sheets.
        b. Pockets.
        c. Veins.
FOREIGN OR TRANSPORTED. —
    A. Sedimentary.
        a. Marine.
            1. Pelagic.
            2. Littoral.
        b. Lacustrine.
        c. Stream.
            1. Flood-plain.
            2. Delta.
    B. Meta-sedimentary.
    C. Residual.
    D. Unassorted.

In this classification, the INDIGENOUS CLAYS, as the term implies, includes all those varieties, commonly called kaolin, which occur *in situ*, or in the immediate place of their origin, where feldspars or other aluminous silicates have been decomposed, and, in part, chemically reconstructed into clays. These are commonly called, by geologists, "residual clays", with the meaning, that they are *left behind*, upon the decomposition of the parent rock. But the process, completed, is one of reconstruction, and a greater percentage of the altering material remains, than is removed in solution. Further, as there are other clay occurrences, which are more properly "residual", in the meaning of the word, the term has been reserved for its apparently more proper use. THE FOREIGN or TRANSPORTED CLAYS include all those, which have been moved, by aqueous or aerial agencies, to points, more or less remote, from the place of their chemical origin. This group of clays differs, among its members, in the manner of their occurrence, and the history of their migration, according to all the varied conditions of their origin and composition (including impurities or accidental elements), the location of their ultimate origin as clays, and the complex relations of transportation and deposition, which may alternate in sequence, and vary in detail, almost indefinitely. Especially is this last true, because of the indefinite term of life, under ordinary conditions, which belongs to the essential minerals of clay, inasmuch as they themselves are the product of the destructive chemical activities, commonly acting on rocks.

The first group under the head of TRANSPORTED CLAYS is, by far, the most important; and it is termed *Sedimentary*, because the clay-beds, classed under it, all have their immediate origin as sediments, as clay minerals, sorted out, in a large measure, by the differential, transportive power of moving water.

The Marine Clays include those, deposited beyond the range of fresh water. They are subdivided into the deep-sea clays, and those bordering the shores.[1]

The Lacustrine Group includes fresh-water lake deposits; and the Stream Group, those clays, which have been deposited, either on the borders, or at the mouth, of the streams, as parts of either flood-plains or deltas.

---

[1] These clay formations, as they are now in process of development, have been exhaustively discussed, in the report on the Deep Sea Deposits, in the reports of the Scientific Results of the Voyage of H. M. S. "Challenger", 1891.

The subdivision, called *Meta-sedimentary*, is meant to include those, which are not transported as clays; but which are the chemical product, from the decomposition of other transported sediments, such as volcanic tufas, pumice etc. That this process takes place, especially in the deep seas, is established by the results of the Challenger expedition.

The subdivision, entitled *Residual*, includes those clays, which have, in their past history, been transported; locked up (in relatively small quantities) in calcareous formations, which have been elevated as land areas, and finally dissolved away, leaving, as a residuary mass, unweatherable clay particles. Our most plastic clays are of this sort, and occur in beds and pockets, in our limestone formations, and in the bottom of limestone caves. The basis of many of our soils is of this immediate origin.

The subdivision, called *Unassorted*, includes those which are commonly called Till or Boulder Clays. They are transported by glacial ice, which has little or no sorting power, and, consequently, deposits them, without definite bedding-planes, and mixed with a heterogeneous collection of sand, pebbles and boulders.

ORIGIN AND COMPOSITION

Clays originate, as has already been stated, as products of the decomposition of various silicates of aluminum. These are chiefly members of the feldspar group, and, probably, are most largely orthoclase and microcline. The feldspars are essentially double silicates of aluminum, and either potassium, calcium, sodium or barium, which often replace each other. With these are usually small percentages of magnesium and iron, due to the presence of mineral inclusions.

The exact state of combination of these elements is not known; and the nature of the change, which takes place, on decomposition, is not completely understood; but, in general, it may be said, that the feldspar molecules are broken up. A portion of the silica combines with the alumina, and these two unite with one or more molecules of water; the alkalies and other elements, present in the feldspar, pass into other combinations, usually soluble salts, according to the nature of the chemical solvent. At, or near, the surface of the earth's crust, it is probable, that this solvent is usually carbonic acid, in which case, the initial accessory products would be largely soluble carbonates.

There are probably other methods, in nature, of decomposing the silicates of aluminum, and producing the essential minerals of clay. Some authors [1] claim, that this has been done, to a large extent, by the action of fluids, containing fluo-silicates, or fluo-borates, which act from below.

Experiments have been made, by Collins,[2] proving, that feldspars are converted into a hydrous silicate of aluminum, by hydrofluoric acid, and indicating, that orthoclase, of the feldspars is the most readily and completely altered, by this reagent, and labradorite, the least so.

The occurrence of secondary minerals, containing fluorine, such as tourmaline, topaz, gilbertite etc., constantly associated with kaolin, the hydrous silicate of aluminum, have led the authors cited, and others, to the conclusion, that hydrofluoric acid has been at

---

[1] Von Buch and Daubrés.

[2] Cf. Min. Mag., Vol. VII, No. 35, p. 205, on "The Nature and Origin of Clays," by J. H. Collins.

least one of the important solvents leading to the production of kaolin.

The ultimate origin of clays is, then, involved in chemical processes, which result, in the production of the essential clay minerals, hydrous silicates of aluminum, associated with other products of chemical changes, and the residuary minerals and mineral fragments, which are undecomposed. Such a mass, if it remains *in situ*, is subject to further chemical changes; to the removal of all or part of the portions, soluble in surface water, and not at first removable in solution; and to the introduction of new material, by infiltrating waters. The additional chemical changes, under ordinary conditions, may consist, either in the continued destruction of the minerals, not wholly decomposed, before the mass becomes essentially clay; or, in the construction of new minerals from the elements, which are already present, or in combination with those, which may subsequently be introduced.

Such clay will vary in character, according to the nature of the rock, from which it is produced; the nature and completeness of the processes; the nature of the chemical solvents; the action of surface water, either in moving or introducing new chemical elements; and, further, according as it may be subjected to the influence of heat or pressure, which tend to convert the clay into slates and other rocks.

All clays have this ultimate origin. The transported clays have, however, a more immediate origin, determined by the conditions of transportation and decomposition, which have already been pointed out. These conditions may vary endlessly; and the endurance of the clay minerals enables them to survive indefinite cycles of removal and decomposition; and they pass from mountains to valleys, from continents to sea-floors, from region to region, until they are finally destroyed, by the metamorphic action of heat.

The hydrous silicates of aluminum, which, according to the definition given above, are the essential minerals in clays, vary in composition, so far as they are modified, by varying amounts of combined water, and different ratios of silica to alumina. Much analytical work has been done, in connection with this problem, with disagreeing results, largely due, probably, to the lack of application, by investigators, of a systematic and uniform method, in making analyses, and to the want of care, by them, in ascertaining, whether the mineral analyzed contained impurities.

Based on the work of various chemists; on that of investigators, using the microscope, with high powers; and on the physical experiments of Le Chatelier;[1] it may now be said to have been established, that there are various hydrous silicates of aluminum, which differ in proportions, in which the elements are combined, and in form and structure, some being colloid or amorphous, and others crystalline, in either the monoclinic or rhombohedral systems. The most important of these is kaolinite, the accepted formula for which is Al_2O_3, 2 SiO_2, H_2O, the percentages being, Silica, 46.3; Alumina, 39.8; and Water, 13.9. Pure clay consists, therefore, of alumina, silica and water, chemically united.

The common mineral impurities are quartz, feldspar, mica, chlorite, pyrite, hematite, limonite, calcite, gypsum, alum, rutile, dolomite, opal, or hydrous silica, and organic matter. These minerals furnish, chemically, in the main, silica (hydrous and anhydrous), potash, soda, lime, magnesia, iron, in various forms, sulphur and titanium. These elements and compounds modify the properties

---

[1] Le Chatelier classifies the hydrous silicates of aluminum, according to the amount of water, chemically combined in them, which he ascertains, by determination of variations in the rate of increase of temperature, brought about by the clay yielding up its combined water, and detected, by means of his thermopyle pyrometer (which has been described in a number of scientific journals), so applied, as to record, photographically, along the scale, the rate of increase of temperature. See Bull. de la Société de Minéralogie, Nos. 5 and 6, 1887, De l'âtcion de la Chaleur sur les Argiles.

of clays, in various ways, both as physical units and by chemical reactions.

The size of the particles, composing the average clay, range from that of coarse sand, relatively small in quantity, and often wholly absent, down to the minutest microscopic particles. Beyond these, are an abundance of fine particles, which will render water opalescent, remaining suspended in it for an indefinite time.

As regards the composition, or mineral identity of the materials, which so remain indefinitely suspended in water, no chemical determinations have been made, so far as is known to the writer. Prof. Whitney includes all these ultra small particles, in what he terms his "clay group." An objection may be raised to this use of the word "clay", which is a term used for rock, consisting essentially of kaolinite. In the case of these minute particles, it has been shown, that they are, in greater part, fragments of this mineral. Theoretical considerations would indicate, that they are likely to be quartz, in as great, if not greater, measure than kaolinite; and, hence, the material might be considered sand, often, with more propriety than clay. The term clay should not be used on the grounds of fineness of material, alone, if this is the case.

The actual nature of these particles being undetermined, their character may be inferred, as follows: —

The great mass of fine grained, fragmental rocks will consist of those minerals, which are both extremely abundant and most highly resistant to the destructional forces, known as "weathering". In these respects, quartz and kaolinite vastly exceed all other minerals.

The rate of abrasion of rock fragments diminishes rapidly, as they decrease in size. A point will, therefore, be reached, where the abrasion of particles suspended in, or transported by water,

THE CLAYS OF GEORGIA

PLATE II

CHARACTERISTIC VIEW IN A RAILWAY CUT THROUGH WHITE CLAY BEDS, NEAR GRISWOLDVILLE, JONES COUNTY, GEORGIA.

will practically cease. Such particles would naturally be in a well rounded condition.

The great mass of particles, smaller than these, will originate in the minute condition, broken from the peripheries of larger bodies. It has been shown by experiment, that minerals, ordinarily insoluble in distilled water, will be found dissolved therein, after abrasion, showing, that blows between masses, large enough for effect, will possibly reach, in tearing substance away, down to the molecule of matter.

From the fact, that the extent of surface increases relatively, as the mass diminishes, small particles are more readily attacked and destroyed by chemical action. So, that the sifting process, tending to preserve quartz and the kaolinite, as residuary materials, is heightened in productiveness.

As to the relative amounts of quartz and kaolinite, among the extremely fine particles, the preference would seem to belong to quartz, unless it can be shown that kaolinite crystals occur with exceedingly small diameters.

Minute products of abrasion will be most abundant, of the material, fragments of which occur in masses, large enough to give and receive, under the conditions of water erosion, effective blows. While this is universally true of quartz, it is only so of kaolinite, to a most limited extent; because the crystals of this mineral, being themselves small, are thus less likely to suffer from abrasion, than is quartz. Despite this fact, however, the presence of a perfect cleavage in the kaolinite needles permits them to suffer most extensive subdivision; but, with this difference from quartz, that, while the latter can furnish an unlimited number of particles, of varying size and shape, the kaolinite particles will rarely be smaller, by natural abrasion, than the cleavage scales, which separate from each other. Owing to the small size of these, they escape further destruction, and preserve, in the main, two of their dimensions intact. This seems generally true; for a microscopic study of clays reveals the essential mineral present, commonly, as small, thin scales, averaging about .003 of a millimeter in diameter, with most of their crystal angles still preserved.

Original size of the fragments and molecular structure, then, seem to limit the smallness of kaolinite particles, and to determine their shape, while they lead to the production of quartz grains of any degree of fineness, and of any shape.

PROPERTIES AND CHARACTERISTICS

The properties and characteristics of clays depend upon, first, the kaolin minerals, and, second, the impurities. In the case of the former, the condition, in which their fragments are found is important; that is, whether their individual particles consist of prismatic crystals, minute plates or thin cleavage scales, resulting from either the simplest form of disintegration of the original crystal, such as is brought about by the action of frost, or from abrasion, suffered, during transportation from its place of origin to its home, as a sedimentary deposit. In the case of the impurities, much depends upon both the absolute and the relative quantities of each, present, and its state of fineness or subdivision.

Clays are characterized by no special *Color*. The pure variety is white; but they commonly occur gray, drab and bluish, often mottled and variegated. They are, also, found of almost all colors, frequently being brilliantly tinted with reds, yellows and purples.

Organic stains give a variety of colors, often brilliant in hue, purple and red being the most striking. Different compounds of iron, the most common coloring causes, give greenish, gray, yellow, brown and red tints. The colors are deepest, when the clay is moist, and they grow lighter, as it is dried.

These colors affect the use of the clay in its raw state; as for example, its use in the manufacture of wall-papers, the adulteration of foods etc.

In most cases, its color, in the natural state, enables us to infer the color of the burned product. Organic stains, however, disappear on burning, while the iron stains are modified or developed, by both the intensity of the heat employed, and the degree of oxidation effected in the process of burning.

Clays are usually characterized by a peculiar *Odor*, or "argillaceous smell", which is not invariable, and by a smooth, unctuous *Feel*, which is also frequently absent, being dependent upon the fineness of grain, and the absence of coarse impurities or grit.

In *Density, Hardness, Tenacity* etc., clays differ widely, being modified, in these particulars, by their geological history. Among the crystalline rocks, where they originate, they are apt to be coarse and loosely coherent; but often transportation and secondary changes, which they may have undergone, following their accumulation as sediments, such as compression, the loss of water, infiltration of mineral matter etc., changes their general structure widely. They vary from soft, porous, incoherent masses, through tough, compact varieties, which are often laminated or have a fissile structure, on to exceedingly tough, and even to varieties, which are so hard, that splinters, flying from beneath the blow of a heavy hammer, will cut the flesh of a workman. Such clays belong to the class, known, commonly, as "flint clays"; and they are mined only by drilling and blasting.

The degree of hardness varies, in the scale of 10, used by mineralogists, from 1 to 3.5; but this is hardness of the clay mass, technically a rock, and not that of the individual clay or kaolinite particles, the hardness of which has not been accurately determined, owing to their extreme smallness.

The specific gravities range, from less than 1 to about 2.5, pure kaolinite having a specific gravity of 2.60. The specific gravity, of course, varies with the state of compactness of the mass. One of the Georgia clays will float, for a time, in water; and, when coated with a thin film of paraffin, it floats like a cork.

BEHAVIOR OF CLAYS, WITH REFERENCE TO WATER AND TO HEAT

The most important properties of clays are those, developed by their behavior, with reference to water and to heat.

Water Absorption, Plasticity, Shrinkage and Consolidation on Drying are the important phenomena, resulting from the relations of clay to water.

The molecules, of which clay and water are composed, have a strong mutual attraction. As a result of this, a single dry kaolinite scale, exposed even to ordinary atmosphere, will rapidly absorb moisture, condensing it, as a film on its surface. This attraction is so strong, that experimental results, by the writer, have shown, that dried clay will absorb moisture, thus gaining in weight, even when enclosed in a chloride-of-lime desiccator.

An aggregation of kaolinite scales, constituting a mass of clay, may be looked upon as a labyrinth of capillary tubes. Owing to the tension, or pull, on the free surface of liquids, which is described by physicists as "surface tension", water will move in small tubes, independently of gravity, falling below its normal position, when there is no attraction between the water and the walls of the tube, to counterbalance this surface tension; as, for instance, if the walls were coated with a film of grease; and it will rise, above its normal height, acting against gravity, when there is an attraction between the water and the substance of the given tubes. These are known as capillary phenomena; and, in the case of clay, with its capillary labyrinths, water will rapidly penetrate the mass, until the supply fails, or, until the weight of water in the tubes exceeds the other operative forces. It is in this manner, that clays absorb moisture, doing so, very rapidly, when the particles are loosely aggregated; and more slowly, though still effectually, when they have been compressed into ever so compact a mass.

On exposing to water, dry clay powder, which has not been compacted by pressure, the former replaces the air in the interstices of the clay; and, owing to the strong attraction between the water and the clay, it draws the particles more closely together, and thus accomplishes a *Primary Shrinkage* of the mass.

Clays, which have not been indurated, by the infiltration of silica, or other soluble mineral matter, or by some form of recrystallization, no matter how dense or hard they may be, will dissolve, when exposed to water, even though the whole mass be not in contact with it. This phenomenon is called "slacking", and is due to the penetration of the water between the minute kaolinite scales, in the manner described above.

If, then, an incoherent, dry clay powder be given an opportunity, it will absorb a definite amount of water, dependent upon the fineness of grain and shape of the particles, and the compactness of the mass, and will shrink, in the ratios, also, of these same conditions.

SHRINKAGE AND CONSOLIDATION ON DRYING

If, in a moistened or wet condition, clay be subjected to heat, but slightly above the boiling point, the moisture will be expelled, or, more properly, will be absorbed by the atmosphere. It departs, under these conditions, molecule by molecule, film after film, being removed from the exposed surfaces, which are those in the open ends of the minute tubes or fissures in the clay. But, as soon as a film of water is removed from one of these tubes, the weight of the column is diminished, and, by "capillary attraction", there is a tendency to elevate the water to its original height. This can only be accomplished, assuming a new supply of water from any source to be prevented, by a possible shrinkage of the whole; or by its breaking into parts, which shrink upon themselves. The completion of this process may be termed, for convenience, the *Secondary Shrinkage* of clay. It results in a more or less consolidated substance, the degree of consolidation depending upon the nature of the clay particles; and, essentially, upon their small size and peculiar shape; that is, that they are minute, thin scales.

These remarkable characteristics, which are phenomena, so common and so apparently simple, that they seem never to have received more than superficial explanation, are produced by forces, brought into play by two physical processes, — nothing more than adding to, and removing water from, clay powder; and, but for the part played by these forces, the material would again be an incoherent powder, at the close of the processes. This aspect of the case seems to have been, hitherto, overlooked by clay investigators, who have explained shrinkage, by the mere statement, that it is due to the loss of interstitial water; but, if this alone took place, unaccompanied by the action of other forces, the wet clay, on drying, would, of course, fall again into its original state. The effect is, that, as the water is removed by evaporation, "surface tension", on the free surfaces of the water, co-operates with the attraction between the molecules of water and kaolinite, and a *pull* is inaugurated, which shrinks on the one hand, and consolidates on the other; the latter, by bringing the clay particles within the realm of mutual attraction, or to the point, where friction, or the interlocking of surfaces, of the particles gives strength to the mass.

The amount of shrinkage, and the degree of consolidation, for clays of a given density, are largely dependent upon the fineness of grain and the shape of the particles, and the condition of the surfaces and edges of these; that is, whether they are relatively smooth or rough.

The coarse clays, or those with much impurity, shrink but little, and show but little tensile strength.

Clays, mixed with enough water, to make them only plastic, shrink in amounts, ranging from about 1 to 10 per cent. When dried, from a condition of self-saturation with water, they shrink in amounts, ranging from 5 to, at least, 30 per cent., linear measurement.

The degree of consolidation, measured in tensile strength, varies enormously, the variation being from a few pounds, in the case of "flint" clays, to between three and four hundred pounds, in the most plastic clays, per square inch. The manner, in which shrinkage and consolidation on drying are brought about, may be explained by reference to the following diagram, fig. 1.

Fig. 1

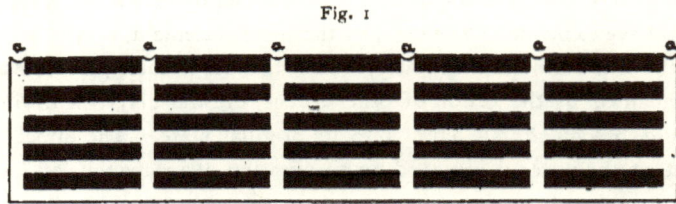

Diagram to Illustrate Shrinkage and Consolidation.

This figure illustrates, diagrammatically, the beginning of evaporation of water from a mass of clay, which has been allowed to absorb the former, to the point of saturation, through perforations in the bottom of the containing vessel, the latter having now been removed from the water-supply.

So long as water could be drawn up from below, while evaporation was going on above, there would be no change in the conditions, except the gradual movement of water upwards; and no necessary movement of the clay particles in any direction.

When the supply of water is removed, surface tension at (a) still operates to elevate water to the original height. If there were no points of weakness in the mass, that is, if the conditions were uniform throughout, and each particle were fixed in its place, there would result, of course, only a gradual subsidence of the water. If the particles, however, be considered free to move in any direction, uniform as regards their size, shape and distribution, and if the affinity of water for the sides of the vessel be exactly equal to that for the clay, a vertical shrinkage will take place.

The capillary attraction, capable, originally, of lifting water to the points (a), would still be operative; and water, to replace that evaporated, would be fed from those channels, the walls of which are the most easily collapsible; and, with the given conditions, these would be the lateral ones, the downward movement being facilitated by gravity and the presence of the one free surface above.

It must be borne in mind, that the tension, at the points (a), and beneath them, throughout the process, is of uniform value in each and every tube.

If, as in the case of our experiment, the affinity of the clay for the water exceeds that of the walls of the vessel, a point will be reached, where the up-pulling at (a), between the clay particles, will remove the water from the columns adjacent to the sides. The mass is then free to shrink upon itself, in horizontal directions, as well as vertical. Water is fed to the exposed surfaces, so long as the particles can approach each other. Shrinkage has been accomplished, and consolidation results, from the bringing of the fine particles, within the realm of mutual attraction, and from inter-surface friction.

When this takes place in nature, the lateral tension is unable to contract the mass as a whole. Owing, however, to inequality of conditions, planes of separation are established along lines of weakness, and so-called "mud-cracks" result. The smaller masses, thus formed, may then shrink farther. In the arts, this phenomenon is called *Checking*.

The finer grained the material, the greater the amount of surface of the particles, as compared with their mass. Consequently, there are more contacts; mutual attraction and friction is more effective; and the mass more completely consolidated. In the case of sand, the mass of the individuals so far exceeds the amount of surface, at which contacts exist, that the bonds are ineffectual, and the material remains incoherent.

The phenomena of shrinking and consolidating, on drying, are all-important points in the economic use of clays, in a great variety of ways. They make possible the sun-dried brick, and most, if not all, of the wares produced from clay, by burning.

In a larger way, these properties of clay effect great results in nature, and some of the facts, in this connection, are worthy of special discussion, from the bearing they have on agricultural problems; and, for this reason, the subject is further amplified. The behavior of clays, with reference to water, is of fundamental importance to vegetation. These properties, in conjunction with "surface tension," operate to retain water at the surface, where it is available to furnish the moisture needed by plant life; and, also, to draw up from the "water-table" below, supplies, to renew that, removed by evaporation and the requirements of vegetation. It carries, held in solution with the water, the mineral foods, which build the tissue, and support the growing plants. It also forms a crust produced through shrinkage, which retards evaporation.

Prof. Whitney, in the introduction to "Some Physical Properties of Soils," states, that chemical analysis has not explained the relation of soils to plants, or the local distribution of the latter; that the general distribution of these is determined by temperature and rainfall, but the local distribution, by the relation of the soils to moisture.

Prof. Whitney (whose works on Soils and Crop Distribution,[1] should be read by every person, interested in scientific farming), in discussing the circulation of water in soils, shows, that it is affected by the presence of salts and organic matter, in solution, and describes the practical results of this, on water and food supplied to

---

[1] "Some Physical Properties of Soils in Their Relation to Moisture and Crop Distribution ";. U. S. Dept. of Agri., Bul. No. 4, 1892. "Conditions in Soils of the Arid Regions ;" U. S. Dept. Agri. Yearbook, 1894. " Reasons for Cultivating the Soil"; Yearbook, U. S. Dept. Agri., 1895.

plants, as modified by the use of fertilizers. Great stress is laid by this author, on the relation of plants to the texture of the soil, which, of course, affects capillary circulation. If the statements, which have been made, above, are true, can it not be reasonably claimed, that, without the properties of clay under consideration, together with the surface tension of water, the very existence of land vegetation, or at least the greater part of it, would be threatened? But for capillarity, would not rain waters escape through or over the soils too rapidly, to be available? Would the film of water, directly attracted by each grain of soil, succeed in resisting evaporation? and, if it did, how extensively would it support vegetation? Where would plant food come from? Another, more fundamental question: — To what extent would the fine-grained materials, constituting soils, accumulate? In short, would land vegetation have a soil, permanent enough to grow in?

From a consideration of the facts, it is evident, that the existence of the former and the permanency of the latter are largely interdependent, and that, besides this, each is directly dependent (the degree being the only question) on the conditions, as stated above.

If, as has been shown, clays and soils should dry without their influence, they would dry into incoherent "powder", "sand" or "dust"; and no crust, more or less solid, would be formed. On the other hand, if they did not operate to elevate, in soils, water from below, they would dry, after wetting, in a much shorter time.

In the former case, soils would be subject to erosion, to a vastly greater extent, than they are, owing to the absence of a crust, and the general consolidation, shown to result from drying.

In the latter case, the presence of the water, obtained by capillarity, renders the soils somewhat tough and coherent, in a different fashion; and so, again, it tends to retard erosive agencies.

Thus, in two direct ways, these materials are preserved from erosion by the wind, the impact of rain, and flowing water.

Indirectly, a growth of vegetation is made possible, which further, and so largely, protects the soil from erosion. Moreover, this vegetation is a source of chemical decomposing agents, which accelerate soil accumulation.

The following generalizations are grouped together, here, as an outline of the features of the circulation of water in soils: — [1]

1st. The soils are composed of mineral matter of different sorts, chiefly quartz, feldspar and clay (kaolin); and these occur in sizes, ranging from that of the particles in coarse gravel down to minute rock fragments, .0001 of a millimeter in diameter.[2] They are, also, of different shapes, the quartz grains varying in form, from angular to well rounded, while the clay or kaolin constituents are commonly present as thin, flat scales.

2nd. The state of aggregation of these, and the relative amounts of each, present in different soils, varies enormously.

3rd. The affinity of the soil particles for water varies with their chemical nature; and particles of the same material have a varying affinity, under different conditions, not, now, well understood.

4th. Salts and organic matter, in solution, modify the value of the surface tension of the liquid, the former generally increasing, the latter decreasing it.

5th. The circulation of water is of two sorts, viz., (a) flow due to hydrostatic pressure, or gravity, and (b) flow due to capillarity.

6th. The permeability of the soil to water, moving under the influence of gravity, depends largely on the size of the tubes. The smaller the tubes, the slower the rate of flow.

7th. The penetration of the soil by water, acting under the influence of capillarity, depends largely, also, but in a reverse way, upon the size of the tube. The smaller the tubes for a given total area of cross-section, the greater the amount of water absorbed, and the higher it is elevated.

8th. For a given soil, the retention of moisture, when loss, from evap-

---

[1] For general discussion, see Whitney, op. cit.

[2] In this connection, Whitney states the interesting facts, that, on the average, about 50 per cent. of the volume of soils is space, occupied by only air and water; that, in a cubic foot of soil, the grains have, on an average, at least 50,000 square feet of surface; and that soils, containing from 10 to 30 per cent. of clay, consist. respectively, of from four to twelve billion grains.

oration, is suffered, depends upon the size of the capillary tubes; the affinity of the soil particles for water; the amount of water-supply; and the length of the capillary tubes, or the distance of the "water-table" below the surface.

9th. Its retention of moisture, when suffering loss through escape below, depends, in a great measure, upon the size of the capillary tubes, the rate of flow, through these, for a given amount of pressure, varying, as the fourth power of the diameters.

10th. Soils, subsoils, and the water reservoir, may all differ in texture. The rate of flow of water to the surface, depends largely upon the nature of the texture of these, and the relative positions, which they occupy. Further, the penetration of the ground, by meteoric waters depends largely upon these same conditions.

11th. Both the presence and circulation of water in soils depend upon very complex conditions, such as, the nature, size and shape of particles, their state of aggregation, or the texture of the mass, and the position and amount of water-supply.

PLASTICITY

Plasticity may be defined, as the capability of a substance for being moulded by a readjustment of its physical units, with reference to the moulding force. Strictly speaking, it is not a property of clay, though it is usually spoken of, as being, perhaps, its most important one. Perfectly dried clay is not plastic. A mixture of clay and water, in certain ratios, which vary within fixed limits, may be extremely plastic. Such a mass may be moulded at will, and readily retains the final shape, to which it is brought.

This behavior is of the greatest use, of course, in manufacturing clay-products, so many of which are moulded, then dried (when they consolidate, as shown above), and finally burned. The cause of plasticity in clays has been a subject of conjecture and discussion, for many years, Cook being one of the earliest American investigators, to call attention to the question. He rightly inferred the relation existing between this property and the shape and size of the individual particles.

A number of observers have assigned plasticity, more or less remotely, to the chemical composition of the clays, — that they are aluminous and hydrous; and this theory, or rather conjecture, is, in a measure, correct. The chemical composition certainly participates in the cause, though, just how it is connected with the phenomenon, the advocates of the theory have not satisfactorily shown.

The fact, that some clays, more impure than others, are the most plastic, has long been known. It has been suggested, that the presence of vermicular-formed crystals, and irregularly angular particles may have a bearing on the question of plasticity. Facts, however, show otherwise; and the suggestion does not deserve to be dignified with the title of "Theory of Plasticity."

It has, also, been shown, by experiment, and has been long recognized, that fineness of grain, and, as Johnson and Blake have brought out,[1] the shape of clay particles are, also, factors; their researches show, that, when these plates occur in aggregates, instead of individuals, the clay is decidedly less plastic. Prof. Haworth has lately made researches along the same line, and has arrived at the same conclusion.

The true explanation of plasticity can best be arrived at, by looking at the conditions involved. For the purpose of these considerations, clays may be conveniently divided into three classes: — *First*, Those, which occur in their place of origin, — that is, the indigenous clays, which consist of plates, variously aggregated, and prismatic crystals; *second*, those, which have been transported, broken up into cleavage scales, and deposited in degrees of fine-

---

[1] On Kaolinite and Pholerite, Am. Jour. Science, II, Vol. XLIII, pp. 35 and 36, 1867; Johnson and Blake.

ness, according to the sorting power of water, moving with different rates of flow; and *third*, those, which have been indurated, either by cementation, recrystallization or some unknown cause, to the point, where they will no longer slack, in water.

In order to make the clays of the last class plastic, they must needs be first ground. The other two classes are both plastic, the former the least so; while, of the latter, those, which are the finest in grain, having been carried the farthest, or deposited in the quietest water, are the most plastic of all. These clays differ, respectively, in a marked degree in plasticity, according to the amount of interstitial water, which they contain. A certain percentage of the latter, differing in the different clays, gives a maximum plasticity.

These facts evidently tend to show, that *Fineness of Grain* is one of the necessary factors; but a comparison of clays with other fine-grained substances shows, that fineness of grain alone is insufficient to account for the facts observed in the behavior of clays.

As already mentioned, the microscopic examinations of Johnson and Blake have shown, conclusively, that *Shape* of the particles is also a determining factor, and theoretical considerations, to be discussed later, accord with this assertion.

Another factor, which is of supreme importance, is the *Mutual Attraction*, already referred to, which exists between water and clay particles, under normal conditions; and it is here, obviously, that the chemical composition of kaolinite has its bearing, though, not in the manner, which has ever been assigned to it, as a cause. Of equal significance, and of necessity as a factor, is the *Surface Tension*,[1] acting on the free surfaces of the water in the clay mass.

---

[1] The term, "surface tension", is the familiar one, used by physicists to denote certain molecular conditions on the superficial film of bodies. Owing to the absence of cohesive attraction on the outside of such films, there exists a relatively greater stability of position among its molecules, with reference to movement outwardly from the mass, and a tendency to diminish the extent of surfaces, or a *surface pull*, — a resultant of molecular forces acting alone from within.

Compared with these factors, the questions of fineness of grain, shape of the particles etc. are insignificant; at least, they are very much less fundamental.

To sum up the situation, we have, in any given suitable sample for demonstration, a mixture of innumerable, minute, thin, flat particles, separated from each other by films of water, the two attracting and holding to each other, with a force, often greater, than that, by which the water attracts itself. The water, which is a fluid, acts as a lubricant, and facilitates the moulding of the mass. For this purpose, enough water must be present, to perfect this property. Having been moulded, the mass has rigidity enough to maintain its shape, when too much water is not present. The difficulty, that seems to need explanation, is, that a heavy mass of particles, mixed with such a liquid substance, can retain almost any shape given to it. The explanation is, that a relative rigidity is given to the water itself by the pull, existing between it and the clay particles, over a vast extent of surface, permeating the whole, and co-operating with *surface tension*, which participates in holding the water, as does, also, *friction*, throughout the mass.

The questions of fineness of grain and shape of the particles become, then, largely, but modifying factors, affecting degree, and being, within large limits at least, modifiers, rather than determinants of plasticity.

For instance, coarse, angular quartz sand, when moistened, becomes plastic, to a considerable extent. It does not, however, on drying, consolidate; but the consolidation of a substance, on drying, and plasticity are, as thus shown, quite different matters, and should not be confused, as they seem to have been; because they are, to some extent, dependent upon similar conditions, and are associated phenomena.

THE CLAYS OF GEORGIA

PLATE III

WHITE CLAY AND SANDS OVERLAID BY TERTIARY ROCKS, RICH HILL, NEAR KNOXVILLE, GEORGIA.

The diminutive size of the particles acts as a factor, in two ways. In the first place, the smaller the particle, other things being equal, the more readily a number of them will move over each other. In the second place, the smaller the particle, the greater the amount of its surface, as compared with its mass. Prof. Whitney, in the papers quoted, frequently calls attention to the function of "clay" particles, resulting from their diminutive size; and he emphasizes the fact, that, as bodies are subdivided, or decreased in size, the extent of surface rapidly increases relatively to the mass. But it should be borne in mind, that the shape of a particle can almost indefinitely extend the relative amount of surface. Thus, for a given mass, angular particles expose more surface than spherical ones. The surface of spherical bodies varies, as the square of the diameter; but, if the particles be flat, the surface can be increased, without diminishing the mass, by change of shape (theoretically), until the molecules of the substance all lie in a single plane. As a matter of fact, mineral particles are generally angular, rather than round, because they result, largely, *primarily*, as minute particles of any shape, rather than being the eroded spheroidal nuclei of larger masses. Thus, in clays, we find the extent of surface at its practical maximum, through minuteness, on the one hand, and shape, on the other.

It is possible, that, if Prof. Whitney, by chance, neglected the *shape*, in his calculations of the amounts of surface exposed in different soils, his determinations being often the greatest diameters of flat particles, the amount of such surfaces would be even greater, than that, indicated by him.

The thin, flat shape, is also best adapted, for giving rigidity to the mass; because the channels between such particles are most uniform in character, and fewer weak spots are possible, where larger capillary tubes, such as would occur among more or less angular particles, would hold an amount of water, in excess of that producing the maximum strength.

One other factor, which should not be overlooked, is in the character of the surface of the particle — that is, whether it is rough or smooth; ragged edges, for instance, such as occur in finely ground clays, tend to diminish plasticity.

The shape of the particle, and the nature of its surface have, also, much to do with the *solidity* of a clay, after drying.

THE BEHAVIOR OF CLAY, WITH REFERENCE TO HEAT

If clay is exposed to a gentle heat, not much above the boiling point, no change will take place, except that of complete drying, unless it shall have been wet enough to cause shrinkage, which, under circumstances, as already indicated, will cause "checking"; and this, particularly, if the heat, even within this limit, is too high, and the process of drying, too rapid. As the temperature is gradually raised, molecules of the water-of-composition, or the combined water, are given off at intervals, until the substance is completely anhydrous. What, if any, molecular re-arrangement of the silica and alumina, remaining, occurs, is unknown; but, finally, the increasing temperature reaches a point, where incipient vitrification takes place. The molecules of adjacent particles pass from the solid to a viscous state, and coalesce, thus uniting the whole into a much more solid and homogeneous mass, than results from mere drying. It is at first porous. But, ultimately, the whole substance fuses, and complete readjustment of the molecules takes place; and the product, as it is ordinarily cooled, is a kind of glass.

These processes, the loss of combined water, and different degrees of vitrifying, produce a third, or *tertiary*, shrinkage of the mass, known as fire-shrinkage. If the clay is free from impurities, an enormous temperature is required, to bring about complete fusion; but, in an electric arc, where the voltage is high, it fuses into liquid glass, more quickly, than wax melts in the flame of a candle. It will also fuse, but less readily, in the oxyhydrogen furnace.

Clay, which is comparatively pure, and fuses only at relatively high temperature, is called *Refractory* or *Fire Clay*.

The processes, by which incipient and higher grades of vitrification are brought about, in the arts, are termed *burning* or *firing*, and are conducted in kilns or furnaces of various sorts.

Impurities of different kinds modify, in different degrees, respectively, the refractoriness of the clay. Two substances, which are, by themselves, highly refractory, will, when in intimate contact, unite chemically to form new compounds, and thus fuse, under these conditions, at comparatively low temperatures. This is the important principle, by which the effect of impurities, in lowering the fusing point in clays, is accomplished.

Besides modifying the fusibility of the clay, the impurities affect the color of the burned product, as spoken of, in discussing the color of clays. The common coloring agent, in burning, is iron, which gives a variety of shades, ranging from buff and salmon, to red, brown and black.

The relative effect, of the different impurities on the fusibility of clay, will be spoken of, under another heading.

USES OF CLAY [1]

Clay is used, in the raw state, chiefly, as sun-dried bricks; for the adulteration of various food and other products; and in the manufacture of paper, principally wall-paper.

For the two latter purposes, fineness of grain, color, and freedom from grit are the essentials.

The wall-paper industry is the most important use, to which clay, in its native state, is put; and it is this demand, which the Middle Georgia clays should seek to supply. Such clay brings from $8.00 to $10.00 per ton, of 2,000 pounds, delivered in New York; or, on the average, about $5.00 per ton, at the mines.

It is used, to add weight to the paper, and to enable it to take clearer impressions, in printing. Originally, a variety of talc and whiting were used; but, at present, the manufacturers use almost wholly the fine-grained, pure-white clays, which slack easily in water, and which are found mostly in New Jersey, Delaware, Maryland, South Carolina, Georgia and Florida.

Fifteen to twenty per cent. of clay is added to each 1,000 pounds of pulp, though the latter usually retains not more than half of this, in its final state.

Not less than 10,000 tons, and probably much more, of this clay is annually consumed, by the wall-paper trade alone.[2]

The use of clay, in the raw state, constitutes but an insignificant part, in the total consumption of this material, in the world. Its chief use is in the manufacturing of burned products, of many

---

[1] See list of publications in appendix.

[2] A brief account of the use of these wall-paper clays will be found in the *Manufacturers Record*, of Aug. 13th, 1897, p. 33, in an article, entitled "Southern Paper Clays," by H. K. Landis.

sorts; and for this purpose, there is manufactured, at the present time, in the United States, alone, a product, valued, probably, at not less than $70,000,000, annually. The statistics, compiled by the United States Geological Survey, show a valuation of our clay-products for 1890, of $67,770,695, and for 1894, of $65,389,784. These figures are, in all probability, too low; as a large number of the minor industries are, unavoidably, omitted, in the reports, which are made to this department.

The burned products may, in a rough way, be classified into household wares, including porcelain and china wares, and art pottery of many kinds; structural materials, including buildings, bridges, culverts, road-beds etc.; drainage materials, such as sewer-pipes and drain-tiles; and refractory materials, that is, products, which will withstand high temperature, and, in cases, the corrosive action of chemical agents. These latter are mainly structural, such as fire-proofing, now used extensively in large city buildings and factories; linings to retorts and furnaces; and utensils, in which glass is melted, fusions made for assays, etc.

Following is a list, including, in a general way, most of the products made of burned clay: —

Common brick; ornamental, pressed or front brick; paving-brick; enameled brick; washed brick; fire-brick; roofing-tile; encaustic, panel and floor tile; terra-cotta; terra-cotta-lumber; drain-tile; sewer-pipe; glass-pots; gas retorts; crucibles; porcelain, common and art pottery; plumbing articles; and electric insulators.

For the methods of manufacturing these products, the manner of preparing the clay, etc., the reader should seek the general literature on the subject, a number of references to which are given, at the end of this report.

GEOGRAPHICAL AND GEOLOGICAL DISTRIBUTION

Clays are common to all countries, having the diverse distribution of fragmental rocks, as well as that of the feldspathic igneous rocks, from which they are being constantly formed.

In the geological scale, clays occur among rocks of all ages; but the indigenous clays are, of course, more recent, than the rocks, among which they occur; and they may have been formed, at any time, subsequent to the solidification and exposure of the latter, from which they are derived. The transported clays occur in all the geological formations, from the Paleozoic age, down to those of the present time; but, so far as they have been preserved in the older formations, they are more apt, to have become indurated by compression, or by the infiltration of mineral cements, or, doubtfully, by recrystallization; and many, of what were once clay-beds, in these formations, have been metamorphosed to slates, by the action of heat and pressure. In the more recent formations, they are less consolidated and more abundant. They are more abundant, here, because, up to a certain limit, the forces, which produce them, are more largely active, than the forces, which destroy them. Further, they are easily and quickly removed, by erosive action, from the older to the newer formations; and they lose nothing, destructively, in essential character, through the process of transportation.

CHAPTER II

METHODS OF SEEKING AND TESTING CLAYS

Clays are located, chiefly by observation in the field; in cuts along highways and railroads; on stream banks; and in gullies, in the fields and forests. They are found, also, in sinking wells, and by boring, and they may be systematically sought by trained geologists, who infer probable location and position, from a variety of observed geological facts.

When found, the expert can arrive at an approximate estimate of their possible utility, by a brief examination, known as macroscopic investigation, this meaning, merely, the hand study of a specimen. The color is observed; the taste and feel are noted; its structure and relative freedom from sandy impurities are determined (by means of a pocket lens, if necessary); and its slacking quality and plasticity are tested, by wetting with water.

Such an expert will, also, note its manner of occurrence; probable thickness and extent; nearness to market, and to means of transportation; the amount of stripping or over-burden, etc.

SAMPLING

As a preliminary to laboratory tests, the clay occurrence should be carefully sampled; as it is useless, to undertake the expensive investigation of a clay, unless it is known, that enough of such material is available to warrant its development. Samples should, therefore, be so collected, as to represent, if not the average of the occurrence, at least a part of it, which can be profitably separated from the rest. This question of sampling is one, of the greatest importance, and it should only be undertaken, by some one, thoroughly familiar with clays and their uses, clay-bodies being very rare indeed, which are uniform in character, for any considerable distance, vertically or areally.

LABORATORY INVESTIGATIONS

For more detailed and more perfect determination of the value of clays, a variety of means may be adopted, for testing the nature of the properties of samples.

PREPARATION FOR TESTING

Since much of the value of laboratory tests depends upon the comparative use of the results, a uniform method should be employed, in preparing the clays for these. It is, of course, necessary to vary the preparation, at times, for special purposes.

In general, clays, which slack, need not be ground; but, that these may be compared with flint-clays, it is sometimes advisable to grind them to a state of maximum fineness, that is, so that all the largest particles shall pass through a sieve of given mesh, which may range from 20 to 100, to the inch, according to the test to be made. As it is necessary to do this, usually, in an iron grinder, it is advisable to pass the clay slowly over an electro-magnet; or, where this is not available, to remove the iron particles, introduced by grinding, by a thorough use of an ordinary magnet.

In order to determine the possible use of some clays, it is necessary to wash them, thereby removing the coarser impurities; and, for some purposes, it is advisable to separate the clay into different parts, according to degrees of fineness. This may be done, by dissolving, physically, the sample in water, in some convenient vessel, agitating, and allowing the coarser particles to subside, and pouring the liquid, containing the finer, into another vessel, a process, which may be repeated indefinitely, ever subdividing the material, according to its fineness.

The following is a brief outline, in a general way, of the usual preparation of clays for laboratory examination, conducted by the writer: —

The sample, which should weigh not less than thirty pounds, is first divided into two nearly equal parts; and, from these, a sufficient supply of specimens is set aside, for the description of the clay, as it occurs in the field; for the purpose of microscopic study; and for the determination of its specific gravity. The first division of the clay is then ground, until it entirely passes through the sieve, of definite mesh. The remaining part is then set aside, as a reserve. From the ground and sifted clay, a part is taken, and dried to constant weight, at 100° C., for plasticity and shrinkage tests. Another part is taken, freed from iron with an electro-magnet, and set aside for the fusibility tests, and to furnish material for chemical analysis.

The laboratory study of the samples then begins, and is made, with reference to their chemical and physical composition; their useful or scientifically interesting properties; and the conditions, on which these properties depend.

SPECIAL METHOD FOR SEPARATION OF CLAYS

In order to separate into classes, according to the shape and size of the particles, those clays, which can be physically disintegrated,

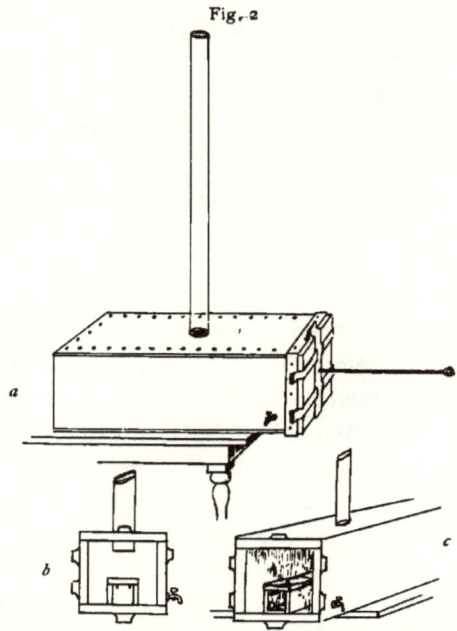

Apparatus for the Physical Separation of Clays.

by very gentle trituration, or which will slack, use is made of an apparatus, illustrated by figure 2. This apparatus was made, as an experiment only [1] ; but it has been found useful. The idea was

---

[1] If the original plan is carried out, an apparatus will be constructed similar to this, but differing in proportions, having a much higher tube, and a longer box, containing a long train of cars, which it is proposed to move, at a regular known rate, by clock-work.

suggested to the writer, by Prof. N. S. Shaler, the plan being, in case of the successful working of this experimental apparatus, to construct one, with such modifications, as would enable it to separate particles, of various shapes and sizes, in clays and soils, and to record, in a measure, the rate of subsidence of such particles, in different solutions.

The apparatus, constructed, consists of a wooden, water-tight, rectangular box, about four and one-half feet long, and six inches in height and width, inside measurement. Inserted in the top of this box, is a glass tube, four feet high and two and one-half inches in diameter [1]. Immediately beneath the lower end of this tube, is a double rail track [2], on which runs a train of zinc cars, connected with the sliding rod shown in *a* of the figure. The bottom of the cars are perforated, and lined, when ready for use, with filter-paper. The front of the box, where the cars are inserted and removed, is made water-tight, by means of a sheet of soft rubber, which is compressed between the box and the end, by six thumb-screw clamps, omitted in the figure. A stop-cock, at a lower level than the cars, drains the water quietly from the box. The sliding-rod moves through a brass cylinder, made water-tight by packing.

In preparing the apparatus for use, the train of cars is inserted, and connected with the piston rod; and the front end is clamped firmly to the box. The apparatus is then filled with water, carefully run in, through a rubber tube, inserted down through the glass one, so as not to wash out the filter papers, which lie in the cars. When all the compressed air in the box has escaped through the pores of the wood, and the level of the water has become stationary, at the top of the tube, the piston-rod is pushed gently in, until the front car is immediately beneath the bottom of the tube. The

---

[1] See fig. 2, *a*.   [2] See fig. 2, *b* and *c*.

piston-rod is marked, so that each car may be brought, successively, into a similar position.

The clay or material, to be separated, is rapidly sifted through a coarse sieve [1] upon the surface of the water, at the top of the tube, at which point it is constantly stirred, by means of a wire rod, having a circular motion, and so used as to prevent, as far as possible, the establishment of vertical currents; or the clay is introduced, already in the wet condition.

The greatest difficulty, in constructing this apparatus, was in getting it absolutely water-tight, under the pressure exerted, when the tube is full of water. A very slight leak will destroy the results; as it is necessary to maintain the column of water, undisturbed, for many hours, in separating the finer clay materials from each other.

The first experiments were conducted, by the use of a mixture of quartz sand, mica and clay, the sand and mica being of two grades of fineness; one, comprising that, which, while it passed through a 30-mesh sieve, was caught on a 40-mesh sieve, and the other, that, which passed through a 60-mesh sieve. The clay was ground, until it had all passed through a 100-mesh sieve. At the first trial, these materials were remarkably well differentiated from each other.

TESTS OF THE BEHAVIOR OF CLAYS WITH REFERENCE TO WATER

Tests, along this line, have been made by the writer, for the purpose of throwing light on the nature and cause of some of the properties of clays. They are continued, in a modified way, for the more exact determination of such properties as have an economic bearing. In order to ascertain the water-absorptive power

---

[1] An ordinary soap-shaker is very suitable.

of clays, and the shrinkage, on drying, of a self-saturated clay, dried samples were taken, which had been ground and passed through a 100-mesh sieve. The method employed was as follows: —

Tin moulds were made, four and one-half inches long, by two and one-half inches wide, by one and one-half inches deep, the bottom being made of perforated tin.[1] These were then filled with clay, by gently pouring it into the mould, obtaining, as nearly as possible, the same densities, which were, however, determined in each case.

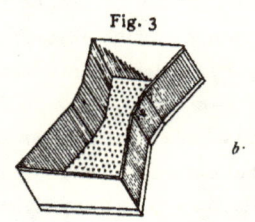

Fig. 3

After obtaining the weight of clay, by difference, the filled moulds were placed in a broad, level-bottom pan, containing water, rising just to the level of the clay in the moulds, a wide pan being necessary, in order to feed to the clays a sufficient amount of water, without materially altering its level.

When the moulds, filled with clay, are exposed to the water, there is a striking difference in the behavior of different samples, some taking much more water than others.

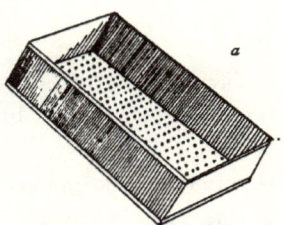

Vessels for Water-absorption Tests, Etc.

The first effect, noticeable, is the down-pull[2] of the clay, the moment it is touched by the water, causing a vertical shrinkage, varying from almost *nihil* to nearly 50 per cent., and varying with the absorptive power of the clay, and with its density. Following

---

[1] See fig. 3.

[2] Clay dried from the wet state, will, however, of course, expand on being again wet, this question depending ultimately on density.

the first down-pull, broad cracks form in the clay, sometimes one-eighth of an inch, or more, across, which are persistent, no matter how much water the clay absorbs; so that it becomes necessary, for further experiment, to slightly stir the whole to a homogeneous mixture. Such cracking is evidently due to the more rapid rising of the water along the sides, whence it penetrates the clay laterally, causing the cracks, by the shrinkage produced. The results of a number of such absorption-tests are grouped in a table, further on. The samples are removed from contact with the water, at the end of twelve hours' exposure, though, in practice, they are found to gain little, if any, after an hour's exposure; the water, adhering to the bottom of the vessel, is removed; the whole is immediately weighed; and the weight of the water, absorbed, is determined, by difference. By use of this method, clays are found to absorb amounts, ranging from 25 to 200 per cent. of their own weight.

SHRINKAGE

The samples, from the above process, are then allowed to air-dry, this being facilitated, by use of the steam-bath, in some cases. When perfectly dry, the shrinkage is determined, by difference, between the inside measurements of the moulds and the dimensions of the resulting brick. Under these circumstances, shrinkages have been noted, running as high as 30 per cent.

For the determination of the shrinkage of clays, on drying, from the practical standpoint, that is, under such conditions as are employed by brickmakers etc., small bricks were moulded, enough water being added, in each case, to make the clay thoroughly plastic, and marks were made near the edges of the brick, and on three sides. The distance between such marks, on the respective faces, were carefully measured, before and after drying, the difference giving the shrinkage. Such shrinkages vary from one to ten, and average about eight per cent. It is impossible, to obtain accurate, comparative results, with this method, because of the difference in obtaining like results, and because of the necessity of using different quantities of water, to obtain the required plasticity, with different kinds of clays.

DETERMINATION OF THE DEGREE OF CONSOLIDATION OF AIR-DRIED CLAY

Since the strength of air-dried clay is an important factor in the manufacture of many clay products, it is deserving of attention among laboratory tests. It may be determined, in a variety of ways. Sticks of dried clay may be suspended by the ends, and the weight, which they will support without breaking, be ascertained. The crushing strength may be determined, as is done in the case of building-stones, bricks etc., by moulding the clay into cubes, and testing them in an ordinary crushing-machine. The resistance to penetration, by a needle or cone-shaped instrument, like Vicat's needle, used in testing cements, may be determined; or the tensile strength may be ascertained, by making brickettes in a mould, similar in shape to that of fig. 3, *b*, and testing with a tensile strength machine; also, of the kind, used in examining hydraulic cements.

METHODS OF DETERMINING PLASTICITY

The capability of being moulded, and of retaining any given shape, possessed by a mixture of clay and water, as has been shown in the description of plasticity, varies, according to a number of factors. For all practical purposes, it can be satisfactorily determined, in a direct way. A few experiments, made in as many minutes, will give the approximate amount of water, required to produce the maximum plasticity, which, it should be remembered, is only the possibility in the substance of internal re-arrangements, sufficient to allow the whole to be modelled, while, at the same time, its rigidity is more than sufficient, to resist collapse or distortions, which gravity tends to produce. Any observer, of common-sense, can thus experiment with clay and arrive, very closely, at its limitations, with reference to its capacity to take and retain, while drying, such shapes as are commonly required in the manufacture of clay products. At the same time, its amount of shrinkage can be closely estimated, and degree of consolidation, noted.

For purposes of comparison, and in the hope of obtaining some accurate and scientific results, experimenters have suggested, from time to time, various methods of procedure. Bischof, who was an eminent German investigator and the author of a work on Fire-clays,[1] made a number of suggestions. One was, that a standard series be adopted, and comparison of other clays made with these, by noting the relative amounts of clay rubbed off on sheets of paper, the rubbing being done with the fingers. Another suggestion was, for a comparison of water-absorptive power of clays, based on the theory, that the more water a clay absorbs, the more plastic it is; and still another suggestion, by this author, was, that comparisons be made, by forcing pencils of the moistened clay samples through a die, in a horizontal position, and noting the rigidity, by measuring the maximum distance, in a horizontal direction, reached by the pencil, from the die.

Another method of expressing the plasticity is, by means of the tensile strength of the dried clay. This was tried by the writer, in the preparation of a Ph. D. thesis, at Harvard, in the winter of 1893–'94.

Of these methods, only two deal directly with the question of

---

[1] Die Feuerfesten Thone. Leipsic, 1876.

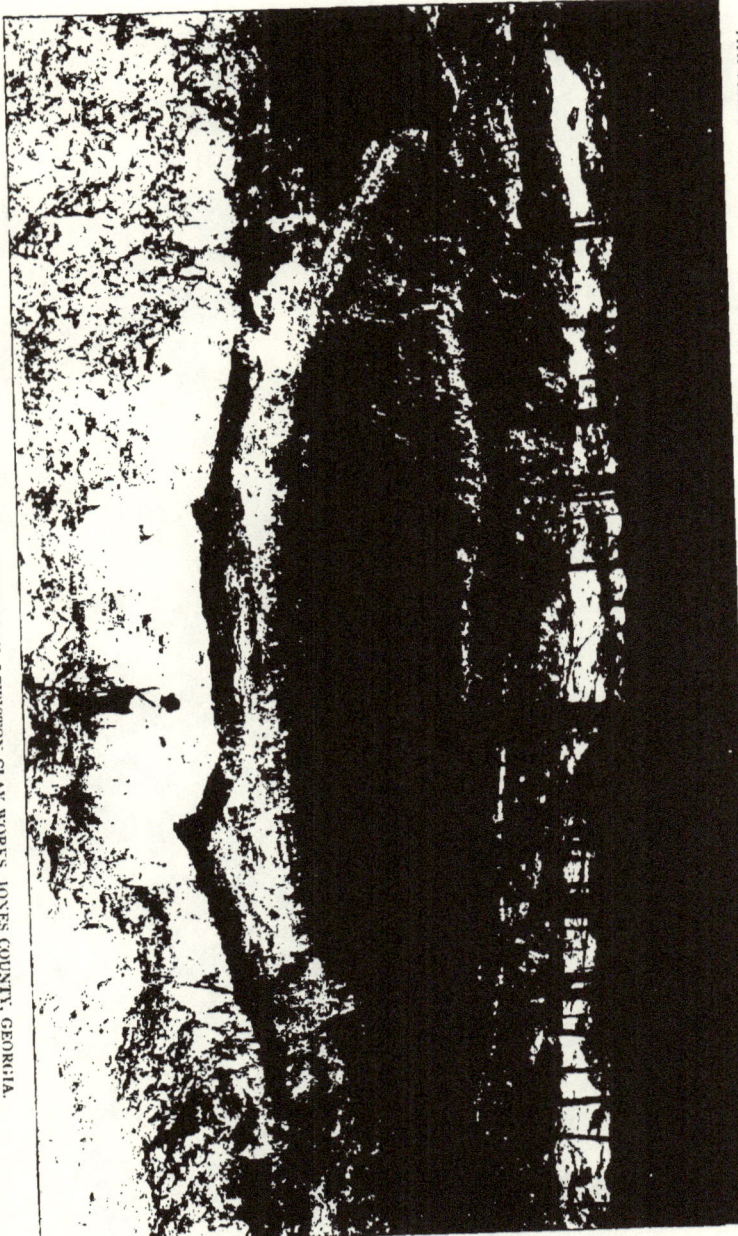

THE CLAYS OF GEORGIA

A SAND BED, UNCONFORMABLY OVERLAYING WHITE CLAY, LEWISTON CLAY WORKS, JONES COUNTY, GEORGIA.

PLATE IV

plasticity; the others are the determinations of properties, essentially different from plasticity, and result, either from the confusion of properties (which seems to be the case), or the assumption, that the variations in two properties are constantly the same; and, if this is the assumption, it does not seem to be justified by facts.

One of the direct methods is that, recommended by Langenbeck,[1] that is, the penetration test, by the use of Vicat's needle. But, in his application of the principle, he assumes, that the greater the plasticity of a clay, the larger the amount of water, needed, to bring it to a definite degree of softness, at which it can be worked. He says: — "In order to determine this point, an apparatus has been devised for pushing with a fixed load a wire rod, or thin-walled cylinder, to a certain depth, within a certain time, into the softened clay.

"The proportion of water, required to soften 100 parts of the dry clay, to the requisite emollescence is taken, as the direct measure of the plasticity."

This method has certain advantages, when confined to the obtaining only of comparative results. But, according to the writer's experience, if employed, as it apparently is by Langenbeck, in the examination of all classes of clays, it is capable of leading to errors. For example, some slightly plastic clays would, when little water is added, be almost as incoherent, as in the state of dry powder and would readily suffer the maximum penetration of the needle. Upon the addition of more water, they would reach their maximum toughness and plasticity; but, upon further addition of water, they would become rapidly softened, or fluid, in nature, and again, be readily penetrated by the needle. There seems no reason, why it should not be employed, for the purpose of determining

---

[1] "Chemistry of Pottery," 1895.

maximum toughness; and it has the advantage of measuring, in a way, both the important factors of plasticity.

The other direct method is that of Bischof, where the mixture of clay and water is pushed through a small cylindrical tube.

This method, like the one just discussed, is capable of giving positive results, the only requirement being, that a number of tests be made, with each clay, using different amounts of water, in order to obtain the maximum plasticity of each.

Of the methods, which have been employed by the writer, that of tensile strength of dried clay, as furnishing an index of plasticity, was abandoned, for reasons which will be explained later; but it has been used, for direct determinations, with wet clays.

Tin moulds were used,[1] which had the perforated tin bottom, and were large enough at the center, to give a one-inch cross-section of wet clay. From this central portion the sides of the mould expand toward the ends, the whole being about three inches in length. The clays are weighed and allowed to absorb water, just as in the case of the absorption tests; they are then dried to different stages of wetness, and submitted to the tensile-strength apparatus.

On account of the lining of filter-paper, the wet clay is readily removed from the mould; but, in most cases, it is too fluid, in the state of saturation, and must first be partially dried, when the diminished cross-section must be taken into account. This method usually permits of the ready determination of maximum plasticity.

An adaptation of the penetration principle has, also, been tried. Experiments have been conducted, by means of the apparatus, illustrated in fig. 4, which consists of balances, suspended by a pulley, so as to be raised or lowered to any required elevation, supporting,

---

[1] See fig. 3, b.

at one end, a pan with weights, and, at the other, a steel-tipped plumb-bob, carefully turned, so that opposite lines, running to the point, include an angle of 40°. The weight-arm is constructed, by a thread, passing through a pulley with an index finger, moving on a scale measuring degrees of a circle.

In some experiments, this plan was modified by attaching a mirror to the weight-pan arm, which reflected a beam of light upon a vertical scale.

The operation is performed, by bringing the clay beneath the plumb-bob, the balance being clamped to prevent swinging. The balance is then lowered, until, while the arms are still horizontal, the point of the plumb-bob just touches the clay, when the balance is unclamped, a weight, is very gently removed from the pan, and the plumb-bob is allowed to settle in the clay for one minute.

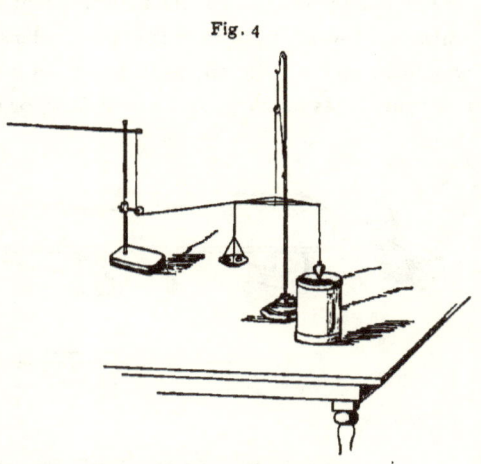

Fig. 4

Apparatus for Use in Plasticity Experiments.

In all the experiments in each of these series, the same weight of clay is taken, and the vessels used, are identical in shape and size. When possible, a 200-gram weight is removed from the pan; but, in the case of the low-strength clays, it is necessary, to remove but 100 grams, and sometimes only 50. The number of grams removed is recorded; and, of course, it should be observed, in comparing the relative strength of the clays.

In addition to these, an original method was tried, by means of an apparatus, devised to test the tensile strength of clays, at different stages of wetness, up to and including that of self-saturation. The requisites for the apparatus were, that it should permit the absorption of water by different clays, under uniform conditions, and should be so constructed, as to grasp, in some way, very wet clay, and prevent, as far as possible, any flow, or change in cross-section,

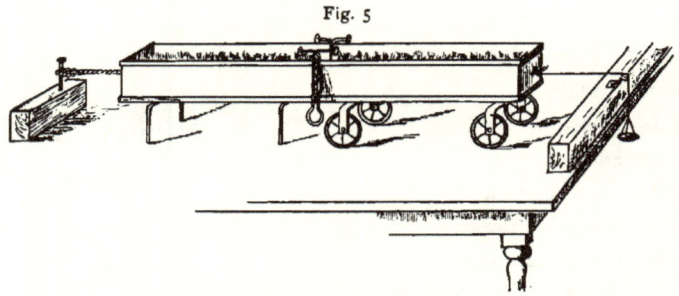

Fig. 5

Apparatus for Testing the Tensile Strength of Wet Clays.

while the breaking test was being made. The apparatus, found to best fulfill these requirements, is illustrated in fig. 5. It consists of two rectangular metal cars, with perforated bottoms, like the moulds above described. These cars have a cross-section, one inch wide and one and one-eighth inches high, and are about four inches, each, in length. The extra depth of the car is to allow approximately for the shrinkage of the clay powder, when wet, so that the final cross-section of the wet clay may be, as nearly as possible, one inch square. On the open ends, are flanges, a quarter of an inch wide, where the cars join. Within each car is soldered a test-tube brush, placed horizontally in the center of the car-box. These brushes are cut off, about a third of an inch from the open end of the car, and are slightly tapered at that end. An end view of one of these cars is shown in fig. 6. One is mounted on clock wheels, so as to run, with the least possible friction, and is connected with a pan, to receive weights, by a cord, which passes over a pulley-wheel, at the level of the center of the car. The other car rests upon wire legs, and is attached firmly to the table, which supports the apparatus.

When ready for use, the open ends of the two cars are brought together, being separated by a thin piece of felt, cut to fit the flanges on the ends of the car, and are held in place by means of clamps. The bottom is lined with filter-paper, to prevent the escape of clay through the perforations.

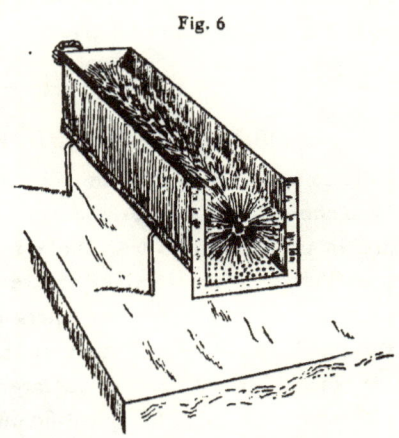

Fig. 6

End View of a Part of the Apparatus for Testing the Tensile Strength of Wet Clays.

The apparatus, thus prepared for use, is filled with dry, powdered clay, gently sifted into the cars, which are jarred sufficiently to insure the complete penetration of the clay among the bristles of the brushes. After weighing, and ascertaining the density, it is placed in a pan of water, so that the surface of the water just reaches the clay through the perforations. When the clay has absorbed its maximum amount of water, the apparatus is returned to its position on the table, the clamps are gently removed, and weights delicately, but rapidly, added, until the wet clay breaks, and the two cars separate.

It is probable, that better results might be obtained, by substituting, for the weight-pan attachment, a spring apparatus, which could be connected with the movable car, and which, on being pulled, until the cars separate, would automatically register the required strength; or, by using the cement tensile-strength apparatus, as in the case of the dried clays.

OBJECTIONS TO THE INDIRECT METHODS

The objections to the indirect methods are, *first*, that it has not been shown, that the variations, in the properties actually tested, are in constant ratio with those in plasticity; *second*, that, in the preparation of the samples for testing, or in the process of conducting the experiment, the "personal equation" is as large, as in the simple, hand examination of the clay-worker; and, *third*, that the results, attainable, have so large a limit of error, as to be insufficiently accurate, for scientific purposes.

The dry tensile-strength test is based upon the assumption, that the plasticity of a clay is due to the interlocking and interlaminating of plates and bundles of prisms; but this involves a misconception of the nature of plasticity, as already shown. Moreover, according to the writer's experience, there is no constant relation, between the properties of the clay, in the dried and wet states.

As an illustration of this fact, experiments, conducted on two different clays, will serve. One of these is from Woodbridge, N. J., and is a soft, "plastic" clay; the other, a sample of the Christy washed pot-clay, from St. Louis, Mo. The accompanying table shows the important facts : —

	Name of Clay	
	Woodbridge, N. J.	Washed Clay, St. Louis
Amount of Clay Base	94.34	56.13
" " Sandy Impurities	1.92	38.32
Percentage of Water Absorbed	75.00	52.00
Tensile-strength Units, in Saturated Condition	60	68
Tensile-strength Units, ⅓ of Absorbed Water Evaporated	330	320
Tensile-strength Units, ⅔ Absorbed Water Evaporated	6,890	1,800
Tensile-strength Units, Dried Condition	4,200	10,300
Shrinkage on Drying	16	19

A study of this table shows the hopelessness of measuring the plasticity of clays, by their tensile strength when dried. What is shown is, that the Christy clay has about 20 times as much sandy impurity, as the Woodbridge clay — in fact, more than one-third of it consists of quartz sand. It absorbs less water; it is less plastic, by the direct-weight tensile process, having a maximum of about one-fourth of that of the Woodbridge clay; and it has a greater shrinkage. But, by the dry tensile test, it is more than twice as strong as the Woodbridge clay. By this method, then, it is more than twice as plastic as the Woodbridge clay, an absurdity and a contradiction of facts. Besides the results shown in the table above, experiments, made with the penetration apparatus, showed, that the ease of penetration of the two clays, the Woodbridge and the Christy, in a state of saturation, is in the ratio of 60 to 57; but, when the same quantity of water is added to each, the amount taken being the average of the amounts required to produce a maximum toughness in each, they showed the ratio of ease of penetrability of 9 to 25; or, in other words, that the Woodbridge clay, under certain similar conditions of wetness, is more than two and one-half times as tough or rigid as the Christy clay.

The last objection to this method, that is, that it is insufficiently accurate, for scientific purposes, is borne out by the admissions of one of its champions, *first*, that a large "personal equation" comes in, in preparing the sample; and *second*, that one, without considerable experience, in employing the method, will get variations in results, of from 25 to 50 per cent. of the strength of each clay; and that the expert can only hope to keep the variation to within 20 per cent.

SUMMARY OF RESULTS

The experiments, conducted with reference to *the behavior of clays toward water*, showed, naturally, results, widely varying among different clays. A brief summary of the more important of them is given below:—

1 Clays absorb water, in amounts, varying, respectively, from 40 to 200 per cent. by weight.

2 Generally speaking, those, which were originally incoherent in nature, absorb the largest amounts of water; and, with little variation, the less the *density* of the dried, pulverized clay, the higher the percentage of absorbed water.

3 The rate of drying varies directly with the amount of water absorbed.

4 The amount of *shrinkage* of a clay varies according to a complex of conditions, such as density, the size and shape of particles

THE CLAYS OF GEORGIA

UNCONFORMITY BETWEEN SANDY CLAYS, IN A CUT ALONG THE GEORGIA RAILROAD, NEAR AUGUSTA, GEORGIA.

PLATE V

·of the clay-base, and their state of aggregation;[1] also, according to the amount of sand present, and the size and shape of its particles.

5 The percentage of secondary shrinkage varies in different directions, being *greatest* for the *smaller dimensions*, the reason for this being, probably, as follows:—

The tendency to establish equilibrium in the mass, due to the loss ·of water on an evaporating surface, is not followed by uniform re- sults, owing to greater friction, and a greater amount of mass, to be moved in the longer direction than in the shorter. Consequently, the movement and shrinkage is more successful, along the shorter dimension. More marked than this, however, as a cause, is the ·difference in the amount of friction, on the contact surface, at the bottom of the brick, along the lateral and longitudinal directions. A slender pencil of clay can be dried, without breaking, only by resting it on a series of rollers, or by some similar device.

6 The tenacity of the clays varies exceedingly, at different stages of wetness.

7 In general, clays, diminishing the most in bulk, through sec- ondary shrinkage, were, in the dried state, the most tenacious. For ·example, one clay, having a maximum shrinkage, in a single direc- tion, of about 24 per cent., withstood a strain, per square inch, ·eleven times greater than another with a maximum shrinkage of less than 8 per cent.

8 Plasticity depends upon the amount of water present; on the shape and size of the particles; and on the attraction of these for water.

---

[1] The kaolinite scales are sometimes grouped together into larger units.

EXPERIMENTS ON THE BEHAVIOR OF CLAYS WITH REFERENCE TO HEAT

The effect of heat on clays is to drive out the combined water; to change the color (when certain impurities are present); to consolidate, and, ultimately, to fuse them.

The experiments, necessary to determine the valuable qualities of a clay, are to burn them to different degrees, noting the color, structure and strength of the burned product, and observing the amount of shrinkage suffered; and to observe the temperatures, required for burning to a desired hardness, and the point, at which fusion takes place. The burning tests are made, very conveniently, in an ordinary muffle furnace. The determination of fusibility, especially for the more refractory clays, requires the use of specially devised apparatus, and some instrument for determining degrees of heat, over as great a range of temperatures as possible, and the means of obtaining comparative results of temperatures, too high to be measured by any instrument.

This recording of the higher temperatures is known as pyrometry, and a great variety of methods have been employed, by different investigators.

A comprehensive study of the measurement of high temperatures will be found in a bulletin of the United States Geological Survey, by Dr. Carl Barus;[1] and the application of various methods to the determination of the fusibility of clays has been presented, in a paper by Hofman and Demond.[2]

---

[1] On the Thermo-Electric Measurement of High Temperatures. U. S. Geol. Survey, Bul. 54, by Carl Barus.

[2] "Refractoriness of Clays," Trans., A. I. M. E., Vol. XXIV, p. 32.

Of the many methods, which have been advocated, the most reliable and convenient, according to the writer's experience, is the

Fig. 7

Illustration Showing How Seger Cones Are Inserted in a Brick-kiln, for Regulating Temperature.

Seger-cone method, and the German modification of Le Chatelier's pyrometer.

This pyrometer consists of a hard porcelain tube about a yard long, within which is a thermo-electric coupling, of platinum and an alloy of platinum and rhodium. This tube is inserted in the fire, where it can be left or removed at will. The coupling is connected, by insulated wires, with a galvanometer, the deflection of which is recorded on a dial, by degrees of temperature. This instrument records temperature, up to 1,600° C.

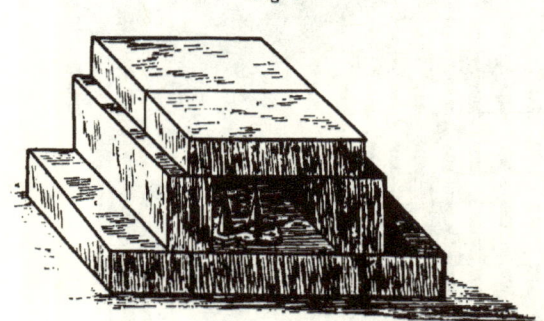

Fig. 8

Illustration Showing How Seger Cones Are Inserted in a Brick-kiln for Regulating Temperature.

The Seger-cone method is one which was proposed and developed by Dr. Seger, of Berlin. He prepared a series of very pointed three-sided pyramids, commonly called "cones," about one inch in height, composed of varying mixtures of alumina and silica, and, except near the top of the series, fluxing bases, such as potash, soda, lime, lead and iron oxides.

The constituents are proportioned, so as to give a difference between successive cones, of approximately 30° C., between cones 022 – 011, and 20° C., between cones 011 – 36.

These cones, shown in Plate XIV, are an extremely convenient and inexpensive method of obtaining approximate melting points; and they are sufficiently accurate, for all practical purposes.

They are extensively used in clay laboratories, and by manufacturers. The manner, in which they are employed, is to expose in a furnace, or kiln, a similar cone of the clay to be tested, together with Seger cones (which are numbered in a continuous series), standing rather far apart in the scale; and repeating the experiment, narrowing down the difference between the standards, until the test specimen melts with one of the latter, or between two, which stand adjacent in the series. Usually, not more than two or three trials are necessary, for a determination.

The manner, in which these cones are inserted in brick kilns, when they are used for the purpose of regulating temperature, is illustrated in figs. 7 and 8, taken from the *Thonindustrie Zeitung*.

Fig. 9 shows, in section, a gas furnace devised to attain a high and equable temperature, for the purpose of obtaining comparative fusing points by this method.[1]

In testing the fire-clays of Middle Georgia, it was found, that the poorest of them have a fusing point, higher than that of platinum, and they are, therefore, beyond the range of study with the Le Chatelier pyrometer. Use was, therefore, made of the upper series of Seger cones; and even these almost failed to give comparative results, the Georgia clays withstanding a temperature, which only the most refractory cone (No. 36 of the Seger series) was able to resist, thus making these clays outrank, in refractoriness, any, which appear to have ever been tested in the United States.

In order to make these tests, it was necessary to use a special furnace, which is manufactured for the purpose, by the "*Chemisches Laboratorium für Thonindustrie*" in Berlin. This furnace

---

[1] This cut is reproduced from Langenbeck's "Chemistry of Pottery," through the kindness of the author and The Chemical Publishing Co., Easton, Penn.

consists of a deep iron cylinder, thickly lined with an extremely refractory mixture of magnesia and pure kaolin. It has a bore of about four and a half inches at the base, narrowing slightly toward the top. The base is a thick plate of iron, perforated with holes arranged in rings, to admit air supplied by bellows, capable of giving a very strong blast. The cones are inserted in a small, covered cylindrical vessel, composed of a mixture, similar to that, which lines the furnace, and having a refractoriness somewhat higher, even, than cone 36. This vessel rests on a solid cylinder of the same shape, size and material, thus being elevated to the point where maximum temperature is attained.

These vessels, before and after using, are shown in Plate XIV.

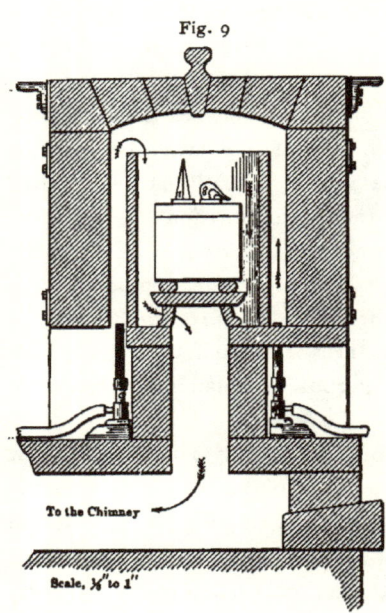

Testing-Furnace for Determining Fusibilities (after Langenbeck).

The following table,[1] from *Thonindustrie Zeitung*, shows the numbers, composition and approximate temperatures, Centigrade, at which the cones fuse:—

---

[1] This table is a copy of one, recently sent the writer by the manufacturers; and it should be approximately correct.

Cone Number	Chemical Composition			Approximate Temperature, C.
022	0.50 Na_2O 0.50 PbO		2.00 SiO_2 1.00 B_2O_3	590
021	0.50 Na_2O 0.50 PbO	0.10 Al_2O_3	2.20 SiO_2 1.00 B_2O_3	620
020	0.50 Na_2O 0.50 PbO	0.20 Al_2O_3	2.40 SiO_2 1.00 B_2O_3	650
019	0.50 Na_2O 0.50 PbO	0.30 Al_2O_3	2.60 SiO_2 1.00 B_2O_3	680
018	0.50 Na_2O 0.50 PbO	0.40 Al_2O_3	2.80 SiO_2 1.00 B_2O_3	710
017	0.50 Na_2O 0.50 PbO	0.50 Al_2O_3	3.00 SiO_2 1.00 B_2O_3	740
016	0.50 Na_2O 0.50 PbO	0.55 Al_2O_3	3.10 SiO_2 1.00 B_2O_3	770
015	0.50 Na_2O 0.50 PbO	0.60 Al_2O_3	3.20 SiO_2 1.00 B_2O_3	800
014	0.50 Na_2O 0.50 PbO	0.65 Al_2O_3	3.30 SiO_2 1.00 B_2O_3	830
013	0.50 Na_2O 0.50 PbO	0.70 Al_2O_3	3.40 SiO_2 1.00 B_2O_3	860
012	0.50 Na_2O 0.50 PbO	0.75 Al_2O_3	3.50 SiO_2 1.00 B_2O_3	890
011	0.50 Na_2O 0.50 PbO	0.80 Al_2O_3	3.60 SiO_2 1.00 B_2O_3	920
010	0.30 K_2O 0.70 CaO	0.20 Fe_2O_3 0.30 Al_2O_3	3.50 SiO_2 0.50 B_2O_3	950
09	0.30 K_2O 0.70 CaO	0.20 Fe_2O_3 0.30 Al_2O_3	3.55 SiO_2 0.45 B_2O_3	970
08	0.30 K_2O 0.70 CaO	0.20 Fe_2O_3 0.30 Al_2O_3	3.60 SiO_2 0.40 B_2O_3	990
07	0.30 K_2O 0.70 CaO	0.20 Fe_2O_3 0.30 Al_2O_3	3.65 SiO_2 0.35 B_2O_3	1,010
06	0.30 K_2O 0.70 CaO	0.20 Fe_2O_3 0.30 Al_2O_3	3.70 SiO_2 0.30 B_2O_3	1,030
05	0.30 K_2O 0.70 CaO	0.20 Fe_2O_3 0.30 Al_2O_3	3.75 SiO_2 0.25 B_2O_3	1,050

Cone Number	Chemical Composition			Approximate Temperature, C.
04	0.30 K_2O 0.70 CaO	0.20 Fe_2O_3 0.30 Al_2O_3	3.80 SiO_2 0.20 B_2O_3	1,070
03	0.30 K_2O 0.70 CaO	0.20 Fe_2O_3 0.30 Al_2O_3	3.85 SiO_2 0.15 B_2O_3	1,090
02	0.30 K_2O 0.70 CaO	0.20 Fe_2O_3 0.30 Al_2O_3	3.90 SiO_2 0.10 B_2O_3	1,110
01	0.30 K_2O 0.70 CaO	0.20 Fe_2O_3 0.30 Al_2O_3	3.95 SiO_2 0.05 B_2O_3	1,130
1	0.30 K_2O 0.70 CaO	0.20 Fe_2O_3 0.30 Al_2O_3	4.00 SiO_2	1,150
2	0.30 K_2O 0.70 CaO	0.10 Fe_2O_3 0.40 Al_2O_3	4.00 SiO_2	1,170
3	0.30 K_2O 0.70 CaO	0.05 Fe_2O_3 0.45 Al_2O_3	4.00 SiO_2	1,190
4	0.30 K_2O 0.70 CaO	0.50 Al_2O_3	4.00 SiO_2	1,210
5	0.30 K_2O 0.70 CaO	0.50 Al_2O_3	5.00 SiO_2	1,230
6	0.30 K_2O 0.70 CaO	0.60 Al_2O_3	6.00 SiO_2	1,250
7	0.30 K_2O 0.70 CaO	0.70 Al_2O_3	7.00 SiO_2	1,270
8	0.30 K_2O 0.70 CaO	0.80 Al_2O_3	8.00 SiO_2	1,290
9	0.30 K_2O 0.70 CaO	0.90 Al_2O_3	9.00 SiO_2	1,310
10	0.30 K_2O 0.70 CaO	1.00 Al_2O_3	10.00 SiO_2	1,330
11	0.30 K_2O 0.70 CaO	1.20 Al_2O_3	12.00 SiO_2	1,350
12	0.30 K_2O 0.70 CaO	1.40 Al_2O_3	14.00 SiO_2	1,370
13	0.30 K_2O 0.70 CaO	1.60 Al_2O_3	16.00 SiO_2	1,390
14	0.30 K_2O 0.70 CaO	1.80 Al_2O_3	18.00 SiO_2	1,410

VIEW SHOWING BOWLDERS OF CLAY, IN THE GRAVEL DEPOSITS, OVERLYING A CLAY-BED IN SOUTH CAROLINA, NEAR AUGUSTA, GEORGIA.

Cone Number	Chemical Composition			Approximate Temperature, C.
15	0.30 K$_2$O / 0.70 CaO	2.10 Al$_2$O$_3$	21.00 SiO$_2$	1,430
16	0.30 K$_2$O / 0.70 CaO	2.40 Al$_2$O$_3$	24.00 SiO$_2$	1,450
17	0.30 K$_2$O / 0.70 CaO	2.70 Al$_2$O$_3$	27.00 SiO$_2$	1,470
18	0.30 K$_2$O / 0.70 CaO	3.10 Al$_2$O$_3$	31.00 SiO$_2$	1,490
19	0.30 K$_2$O / 0.70 CaO	3.50 Al$_2$O$_3$	35.00 SiO$_2$	1,510
20	0.30 K$_2$O / 0.70 CaO	3.90 Al$_2$O$_3$	39.00 SiO$_2$	1,530
21	0.30 K$_2$O / 0.70 CaO	4.40 Al$_2$O$_3$	44.00 SiO$_2$	1,550
22	0.30 K$_2$O / 0.70 CaO	4.90 Al$_2$O$_3$	49.00 SiO$_2$	1,570
23	0.30 K$_2$O / 0.70 CaO	5.40 Al$_2$O$_3$	54.00 SiO$_2$	1,590
24	0.30 K$_2$O / 0.70 CaO	6.00 Al$_2$O$_3$	60.00 SiO$_2$	1,610
25	0.30 K$_2$O / 0.70 CaO	6.60 Al$_2$O$_3$	66.00 SiO$_2$	1,630
26	0.30 K$_2$O / 0.70 CaO	7.20 Al$_2$O$_3$	72.00 SiO$_2$	1,650
27	0.30 K$_2$O / 0.70 CaO	20.00 Al$_2$O$_3$	200.00 SiO$_2$	1,670
28		Al$_2$O$_3$	10.00 SiO$_2$	1,690
29		Al$_2$O$_3$	8.00 SiO$_2$	1,710
30		Al$_2$O$_3$	6.00 SiO$_2$	1,730
31		Al$_2$O$_3$	5.00 SiO$_2$	1,750
32		Al$_2$O$_3$	4.00 SiO$_2$	1,770
33		Al$_2$O$_3$	3.00 SiO$_2$	1,790
34		Al$_2$O$_3$	2.50 SiO$_2$	1,810
35		Al$_2$O$_3$	2.00 SiO$_2$	1,830
36		Al$_2$O$_3$	1.50 SiO$_2$	1,850

NEW EXPERIMENTS

Some time ago, the writer made a number of experiments, in connection with the fusibility of fire-clays, with a view to determining a method, which would give comparative results, in the simplest and quickest way. To this end, attention was turned to the possibilities of the electric furnace and the oxy-hydrogen blast-furnace.

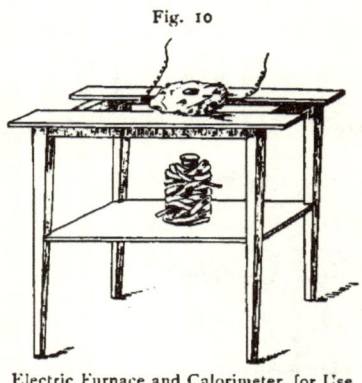

Fig. 10

Electric Furnace and Calorimeter, for Use in Fusibility Experiments.

THE ELECTRIC FURNACE

An illustration of the electric furnace used in these experiments is given in fig. 10. The furnace consists simply of a block of lime,[1] having two holes drilled through it, intersecting each other at the center, at right angles, the horizontal hole being the size of common electric-light carbons, and the vertical hole varying from 1 x 1 inch to 1 x 2 inches, according to the requirements of the experiment. For some experiments, the vertical hole does not go entirely through the lime. The illustration shows a calorimeter in position beneath the furnace; but, in actual experiment, this was arranged, to be raised to, or lowered from, the furnace. The electric carbons, connected with an Edison dynamo, were inserted in the fur-

---

[1] These lime furnaces are readily made in five minutes time, on a turning-lathe.

nace through the horizontal holes, the points coming in contact, at the intersection with the vertical one, the arc being drawn out, when needed. The current, used to establish a suitable arc, of an inch to an inch and a half, registered 140 volts and 40 amperes.

THE OXY-HYDROGEN FURNACE

The oxy-hydrogen furnace used consisted simply of a block of lime, in the side of which a hole was drilled, for the admission of the oxy-hydrogen flame, and into the center of which, from the top, was another hole, for the admission of the clay to be tested. The furnace was hollowed out, in order to provide against contact of the clay with the lime.

The gases for this furnace were supplied from tanks, where they were each kept under the same hydrostatic pressure.

CLAY PENCILS

The clays, to be tested in these furnaces, were prepared as follows: — The clay, powdered to pass through a 100-mesh sieve, had all the magnetic iron removed by means of an electro-magnet; it was then made into prisms about fifteen centimeters long, with a basal section, 1 x 1 centimeter. These are the dimensions of the wet pencils; but, when ready for use, they vary from this size, in proportion to their amount of shrinkage, on drying and burning.

The pencils were made in a mould, lined with filter-paper, the clay being wet with sufficient water, to make it plastic, and pressed into the mould with a spatula. The sides of the mould were then removed, the pencils lifted by means of the filter-paper, and put away to dry. Of several methods tried, this proved the most effective, and gave the largest percentage of unbroken pencils.

USE OF THE ELECTRIC FURNACE

ATTEMPT AT TEMPERATURE MEASUREMENTS

The electric furnace was used mainly for preliminary experiment, the original aim having been, not only to get comparative results, but to ascertain, to a reasonably close approximation, the absolute temperature of fusion for the various clays experimented on. The difficulties to overcome, in order to make possible any satisfactory results of this sort, were, however, too many. The method undertaken was to insert the clay pencil in the furnace, from above, through the vertical opening, avoiding contact with the lime, and to fuse the clay off, in drops, which fall into the calorimeter below. Experiment showed, that the size of the drops was nearly uniform, for a given clay; and it was thought to avoid the difficulty presented by the unknown factor, the latent heat of fusion, by placing the calorimeter, so that the surface of the contained water would intercept the falling drop at the exact point of resolidification. It was found, that such a point could be located, by means of the different effects on the surface of the resulting glass bead, in accordance with its condition, either liquid or solid, on reaching the water. Up to this point, the preliminary experiments promised much. By marking the position for the calorimeter, on its support, when all was in readiness for the experiment, a final reading could be taken, of the temperature of the contained water, and it could be instantly placed in position, the pencil inserted in the furnace, and, in half a minute more, a sufficient number of drops fused off, to raise the temperature of the water in the calorimeter, a desirable amount, for best results. The glass beads, which are exceedingly tough and elastic, when cold, usually remained unbroken, on striking the water; and, after the final temperature-reading was made, they could be taken out and weighed. The water, of course, was weighed as a preliminary.

A number of these experiments were made, with fairly good results, assuming that the value 0.35 expressed the specific heat of the glass bead, and that it was constant, at different temperatures.

The specific heat of substances, however, varies at different temperatures; and the rate of variation is not known to be constant with increasing temperature. It is probable, that, while it increases with the temperature for some substances, it may decrease with others. Very little experimenting seems to have been done on this subject, especially with high temperatures. The writer hoped to make a series of determinations of this specific heat of fused fire-clay glass, at a range of temperature, from 200° C., up to the melting point of the substance.

Since it was not possible, owing to lack of time, to verify, by duplicate experiments, the determinations made, and, further, since it was a first trial, and the method was comparatively crude, the results are worthy only of being recorded as experiments.

For these, a number of glass beads were fused off clay pencils in the electric furnace, and placed in a large platinum crucible, which was exposed to the heat of a blast lamp, run by power, for a half-hour, for each experiment, the temperature not being taken, until it was approximately constant. The beads weighed, on an average, not over half a gram; so they were easily heated through, within the given time. The wire thermopile was inserted in the crucible, and brought in contact with the glass beads, used in the experiment. It was not allowed to touch the bottom of the crucible. The calorimeter used was a small brass cylindrical vessel, within a similar larger one, and separated from it, by cotton-wool packing.

The data of four experiments are given below: —

Experiment A

Number of fire-clay glass beads	3
Weight " " " "	1.3144 grams
Temperature " " "	461° C.
Weight of water	75 grams
Temperature of water	21° C.
" " mixture	24° C.
Water equivalent, of calorimeter	4.7
Calculated specific heat	.416

Experiment B

Number of fire-clay glass beads	3
Weight " " " "	1.3144 grams
Temperature " " "	506° C.
Weight of water	75 grams
Temperature of water	20.5° C.
" " mixture	23.9° C.
Water equivalent, of calorimeter	4.7
Calculated specific heat	.427

Experiment C

Number of fire-clay glass beads	3
Weight " " " "	1.3144 grams
Temperature " " "	712° C.
Weight of water	75 grams
Temperature of water	23.0° C.
" " mixture	26.9° C.
Water equivalent	4.7
Calculated specific heat	.343

Experiment D

Number of fire-clay glass beads	2
Weight " " " "	.8509 grams
Temperature " " "	927° C.
Weight of water	75 grams
Temperature of water	23.2° C.
" " mixture	26.0° C.
Water equivalent	4.7
Calculated specific heat	.366

Grouped Results

A	Temperature	461° C.	Specific Heat	.416
B	"	506° C.	" "	.427
C	"	712° C.	" "	.343
D	"	927° C.	" "	.366

In the last experiment, one of the three beads heated stuck to the crucible, and did not roll into the calorimeter, showing incipient melting. Another experiment failed, in which the beads were heated to 1,100° C., owing to their incipient melting and adherence to the crucible.

Some of the difficulties encountered may be easily obviated, and a sufficient number of determinations may be made to give average results of practical value, at least in a comparative way.

Before making these determinations, 0.35 was arbitrarily assumed as the specific heat, at the fusing point of fire-clay; and, on this basis, a number of pencils, some of which were mixed with calcium carbonate, in amounts varying from 1 to 4 per cent., were fused; and they gave as results, by the calorimeter method, temperatures of fusion, ranging from 1,990° C., to 2,651° C., the latter being the result obtained, on fusing the unadulterated standard clay. As before stated, these are recorded, only as experiments, and not as indicating undoubted temperatures of fusion. The results are too high, owing to super-heating of the drops.

COMPARATIVE FUSIBILITY

An attempt was made, in several ways, to utilize the electric furnace, to obtain comparative fusibilities. One method was to insert the clay pencils continuously, as fast as drops were fused off from the lower end, and to take the average duration of time between the falling of drops, as a basis of comparison. This method, however, failed to differentiate clays, whose fusibility points were near each other.

Another method tried was to insert in the furnace, in a similar way, for the same length of time (a few seconds), one pencil after another, with an occasional repetition of a standard pencil, as a check on the temperature. This method gave unsatisfactory results, owing to the ease, with which the arc changes its position.

Experiments were also tried, for getting a comparison between two clays, at one time, by inserting a pencil of each, firmly clamped together, about a quarter of an inch apart, so as to allow the arc to run between them; but, owing to the very slight vibrations of the latter in the air currents, very different amounts of fusion were produced on the pencils of the same clay; so, that the method failed.

The last experiment, tried with the electric furnace,[1] was to make pencils, as before, inserting, near one end, completely covered

---

[1] It is possible to use the electric arc, without the furnace; but it is much less satisfactory, the arc being frequently blown out by currents of air, and readily cut off by the insertion of the clay pencil, which, apparently, utterly refuses to conduct the electricity; whereas, in the highly heated furnace, the arc readily maintains itself, gliding from one side to the other, of the pencil. There is much less danger, also, in using the furnace, there being a better opportunity to protect the eyes

THE CLAYS OF GEORGIA

PLATE VII

UNCONFORMITY BETWEEN A MANTLE OF LAFAYETTE SAND, AND FIRE-CLAY BEDS, AT STEVENS' POTTERY, BALDWIN COUNTY, GEORGIA.

with the clay, bits of metals and alloys, where possible, in the form of wire, in which case they were inserted parallel with the length of the pencil. These were exposed to the heat of the arc, until the ends were fused to glass, about a third of an inch up the pencil; then they were broken open, and a study was made of the effect on the enclosed metals, among others, platinum, iridium and osmium being tried. But this method was given only a preliminary and incomplete trial.

USE OF THE OXY-HYDROGEN FURNACE

The oxy-hydrogen furnace was used, in conjunction with the electric furnace; and, for obtaining comparative results, it was much more satisfactory, on account of the lower heat developed, which, however, was ample to fuse the most refractory clay.

The comparative fusibility of the selected clays was readily determined in this furnace. The attempts, however, to compare them in a series, was only partially successful, on account of unavoidable changes in temperature, during the appreciable time consumed, in making a number of fusions. The method, giving the best results, was, to compare the clays, each with each of the others, repeating these comparisons, and varying the time of exposure, to make the results more conclusive.

The experiment was conducted, in practice, as follows:— The

---

and face from the intense light. The writer had several unpleasant experiences from this source of danger, being made ill, as was also Mr. Adams, Professor in Electric Engineering in the Lawrence Scientific School, owing to insufficient protection from the light; the skin, which had been exposed, becoming sore, as from a severe sunburn. The eyes, brain and nervous system were affected for more than a week, the symptoms being disagreeable, if not alarming.

hydrogen being lighted, the oxygen turned in, and the flow of gases regulated, to provide the required temperature, the two marked pencils to be compared were inserted through the vertical opening, into the furnace, one after the other, by means of tongs, care being taken to lower each the same distance in the furnace. By using dark-colored glasses, the effect of the heat on the clay could be watched, throughout the experiment. The time of exposure for the first comparison was usually ten seconds; for the second, fifteen; for the third, twenty; and for the fourth, thirty; but these periods were made to vary, according to the temperature obtained in the furnace. At the end of each comparison, the ends of the pencils were broken off, labelled and laid aside, for subsequent examination, unaffected portions being used in the next trial.

Given the oxy-hydrogen plant, this seems to furnish a direct, practical, rapid and simple method of obtaining comparative fusibilities of fire-clays. A set of standard pencils can be made, in accordance with a scale of fusibility; and, by comparisons, through this method, the refractoriness of any clay can be obtained, in a satisfactory manner, in less than five minutes. It involves vastly less labor and time, than the Seger-cone process, conducted as described above, and is equally accurate in results. The lime furnace, after use, can be removed to a desiccator, and kept indefinitely.

INDIRECT METHODS

Most of the indirect methods, for determining the fusibility of clays, are attempts to express the refractoriness, by mathematical formulæ, which take into consideration the chemical and physical constitution of the clays.

The formula, advocated by Bischof, for determining what he calls the "refractory quotient," is based, wholly, on the chemical composition, and is the result of the division of the quotient of the oxygen of the fluxes into the oxygen of the alumina, by the quotient of the oxygen of the alumina into the oxygen of the silica:—

$$\frac{\text{O in Al}_2\text{O}_3}{\text{O in RO}} \div \frac{\text{O in SiO}_2}{\text{O in Al}_2\text{O}_3}$$

Seger recommends modifying the relation of the silica and alumina, by adding the ratio of the fluxing constituents to the alumina, with that to the silica, and multiplying the same by the quotient of the latter into the former, as follows:—

$$\frac{\text{O in Al}_2\text{O}_3}{\text{O in RO}} + \frac{\text{O in SiO}_2}{\text{O in RO}} \times \frac{\text{O in Al}_2\text{O}_3}{\text{O in RO}} \div \frac{\text{O in SiO}_2}{\text{O in RO}}$$

While these formulæ are to some extent useful, they do not take into account the effect, which the physical condition has on refractoriness, the point, to which attention was called by Prof. Cook in 1878.[1]

Wheeler suggests a formula, for taking these physical factors into consideration; but it has not been generally shown to be practicable. It is given in his own language, as follows:—[2]

"$F.F. = \dfrac{N}{D + D'}$, (A), when the clays have the same specific gravity and fineness of grain.

In this formula, F.F. represents numerical value of the refractoriness. N represents the sum of the non-detrimental constituents, or the total silica, alumina, titanic acid, water, moisture and carbonic acid. D represents the sum of the fluxing impurities, or the alkalies, oxide of iron, lime and magnesia. D' represents the sum of alkalies, which are estimated to have double the fluxing value of the other detrimentals, and hence they are added twice.

This formula is found to give fairly good comparative values of the refractoriness of clay, that do not differ more than 0.2 from one another in density (the closer the specific gravity the more reliable the comparison), and are

---

[1] "Clays of New Jersey," Geol. Surv. of N. J., 1878, by Cook and Smock.
[2] Missouri Geol. Surv., Vol. XI, "Clay Deposits," by H. A. Wheeler, 1896.

of similar fineness of grain. When the clays to be compared differ in density and fineness, it is necessary to modify formula (A) by a constant C, that will have different values depending on the density and fineness, so that the formula will be:—

$$F.\,F. = \frac{N}{D + D' + C}$$ (B), in which N, D and D' will have the same values as in (A).

> C = 1, when clay is coarse-grained, and specific gravity exceeds 2.25.
> C = 2, when clay is coarse-grained, and specific gravity ranges from 2.00 to 2.25.
> C = 3, when clay is coarse-grained, and specific gravity ranges from 1.75 to 2.00.
> C = 2, when clay is fine-grained, and specific gravity is over 2.25.
> C = 3, when clay is fine-grained, and specific gravity is from 2.00 to 2.25.
> C = 4, when clay is fine-grained, and specific gravity ranges from 1.75 to 2.25.
>
> These values of C are only approximate."

For determinations of refractoriness by this method, the long and tedious process of a complete quantitative chemical analysis, which should be made in duplicate, must be gone through, and then the specific gravity and fineness of grain, determined.

The trouble with all these methods, even if they were known to give accurate results, is, that they are too expensive and troublesome to employ.

So long as we can obtain, in an hour or two's time, at the outside, the relative fusibilities, and, in most cases, the actual fusing points, the use of these formulæ is as unnecessary as it is complicated and expensive; and the results are unreliable.

As opposed to these methods, which seek to deduce the relative refractoriness of clays, by formulæ based essentially on the chemical composition and densities of the clays, are the experimental methods, classified as the direct and indirect, which have been thoroughly discussed in the paper, by Hofman and Demond, quoted above, the various modifications of methods being assigned to the originators, Knaffl, Bischof, Otto, Seger and Cramer.

The direct experimental methods are those, which seek to compare the effects produced, of different clays, by exposure to the same temperature. In some methods, as Knaffl's, the point of a small cone of clay is exposed to the blowpipe flame, and the result is compared with that of a similar experiment on another clay. Otto burns three test-bricks, placed alternately with two standard-bricks, in a coke, forced-draught furnace, and compares the test- and standard-bricks respectively (with each other), to ascertain the uniformity of temperature, obtaining the relative refractoriness by comparison of the test-brick with the standard. Seger and Cramer have made cones of artificial mixtures, having a wide range of refractoriness, which are used as standards.[1]

The indirect experimental methods are based on the principal of either toning up low-grade clays, by the addition of definite amounts of more refractory materials, or the toning down of high-grade clays, with definite amounts of fluxes, until the clays have similar refractory qualities, at the same temperature.

An arithmetical expression of the results is obtained, by a simple calculation, based on the amount of material added to the clay. These methods involve the necessity of reproducing the same temperature, for different experiments. To overcome this difficulty, Hofman and Demond devised a furnace, especially constructed to maintain constant temperatures over a considerable area in the furnace. As a standard of measurement, they adopted Le Chatelier's pyrometer; and they used, as a flux, calcium carbonate diluted with silica, their idea being, to so lower the fusion points of clay, as to make comparisons possible, at low temperatures, which are obtained in a more manageable way, and to permit, at any time, the observation of the effects of the heat on the samples tested.

The results of the experiments of these gentlemen, as published

---

[1] See p. 59.

by them, promised much for the method; but later experiments failed to prove its usefulness. It apparently worked nicely, on making up the mixtures, fusing them, and measuring the temperatures; but, when a definite temperature was assumed, and the attempt was made to prepare mixtures, that would fuse at that temperature, the behavior of the mixtures was not characteristic enough, to make it the basis of a test.

CHEMICAL ANALYSIS

The chemical analysis of a clay, if unnecessary and inadequate, as a means of determining refractoriness, is frequently of use, technically, for the purpose of locating the cause of certain good or bad properties, the latter of which may be often remedied by special treatment.

Of a number of methods, which have been recommended, the following has been found, by the writer, the most satisfactory in the way of results, and the simplest in practice: —

Three grams of the finely powdered clay are dried, for several days, at 100° C. to a constant weight, and the loss is calculated, as *Hygroscopic Moisture*. The analysis proper is then calculated, on the basis of this dried clay. The clay, from the above determination, is ignited, at an intense red heat, to a constant weight, and the loss is calculated, as *Loss on Ignition*, or *Combined Water, Carbon Di-oxide* and *Organic Matter*. The clay is then fused with sodic carbonate; and the resulting cake is dissolved in water acidulated with hydrochloric acid; evaporated to dryness; and moistened with hydrochloric acid, which is diluted with hot water, after standing for half an hour. The fluid is decanted on a filter; and, after a repetition of this process three times, the precipitate is transferred to the filter, washed, ignited and weighed as the total *Silica, SiO$_2$*. The residue from the SiO$_2$, after treatment with hydrofluoric acid, is weighed, and set aside, for separate analysis. It is, probably, mostly titanium, which has not been determined in the Georgia clays, thus far analyzed. The filtrate and wash-water is divided into two equal parts, in each of which the iron, calcium and magnesium is determined. In the first of these two parts, the *Iron* and *Alumina* are precipitated with ammonic hydrate; this is washed, first by decantation, and finally on the filter paper, until free from chlorine; and it is then dried, ignited and weighed.

After concentration of the filtrate from the alumina and iron, the *Calcium* is precipitated by ammonic oxylate, which is filtered, washed, ignited and weighed as *Lime*. From the filtrate of the calcium determination, the *Magnesia* is precipitated, and weighed as *Magnesia Pyro-phosphate*. After weighing the ignited alumina and iron, it is fused with acid potassic sulphate, and the iron is determined (after reduction with zinc) by titration with potassic permanganate.

In the second filtrate, the iron and alumina are precipitated and washed; dissolved in sulphuric acid; freed from hydrochloric acid by evaporation; and the iron is then determined by titration, as before. The calcium and the magnesium are then precipitated as the oxylate and pyro-phosphate respectively, and weighed as before.

The alkalies are determined, from new portions of clay, by Prof. J. Lawrence Smith's method, as given in Fresenius.

The sand is then determined, by digesting a portion of clay, for twelve hours, in sulphuric acid, and evaporating to dryness; boiling with sodic carbonate, and weighing the washed and ignited insoluble residue as sand.

Before the final grinding of the clay for analysis, all the magnetic iron is removed with a magnet.

CHAPTER III

THE FALL LINE CLAYS

INTRODUCTORY REMARKS

Beginning a study of the Clays of Georgia, in the Spring of 1896, the writer first turned his attention to the vast extent of stratified deposits in the Southern half of the State. The following notes embody the results of several months' field-work in South Georgia. They contain, also, the results of the chemical analyses and physical tests, made on the Clays of the Fall Line region of South Georgia. They have been accumulated, in the intervals between field-work on the Clays and general laboratory work, the writer acting for the Survey, at the time, in the capacity of both geologist and chemist.

The field-work in South Georgia was begun, by trips down the large rivers, — the Chattahoochee, Ocmulgee, Altamaha and Savannah — the starting points being Columbus, Macon and Augusta. It was on these excursions, that the white kaolins, occurring along the Fall Line, were encountered. Later, much time was spent in tracing them across the State, and in determining their geological horizon. This task was a difficult one, owing to the close similarity in the materials of strata belonging to the different geological periods, and to complicated unconformities between these strata.

The chief results of the work done are: — First, the tracing of

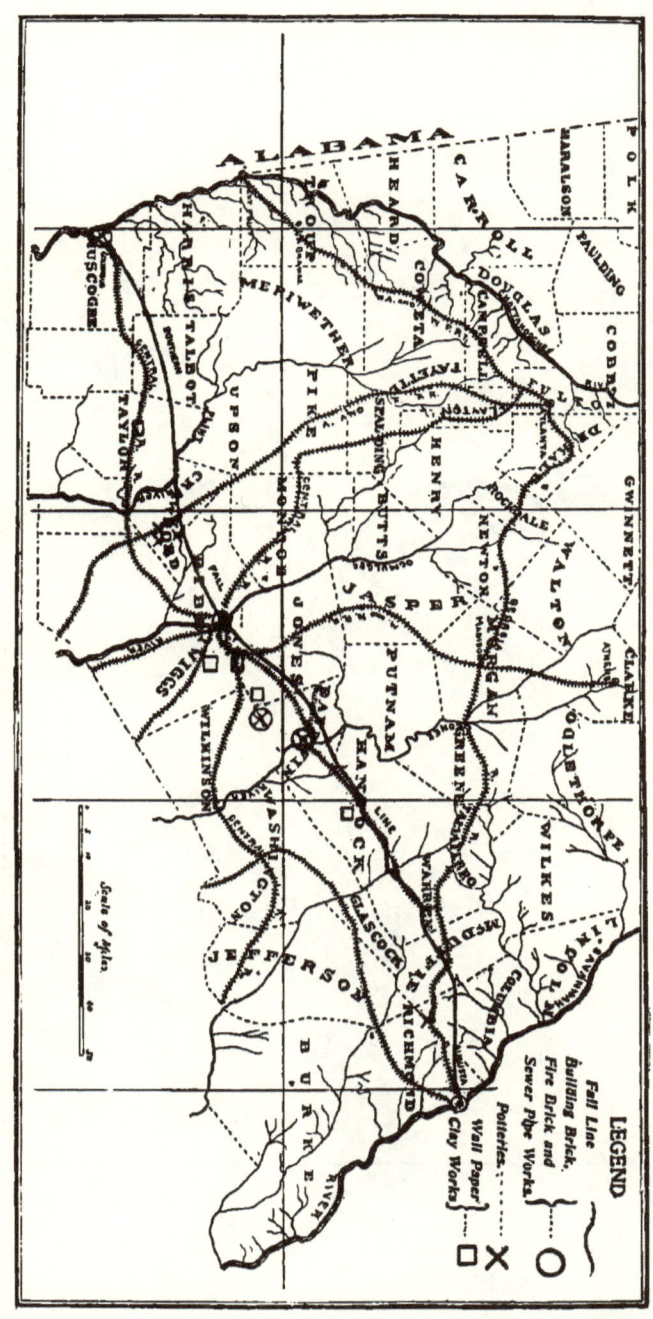

Map Indicating the Clay Deposits and the Location of the Clay Industries along the Southern Fall-Line in Georgia.

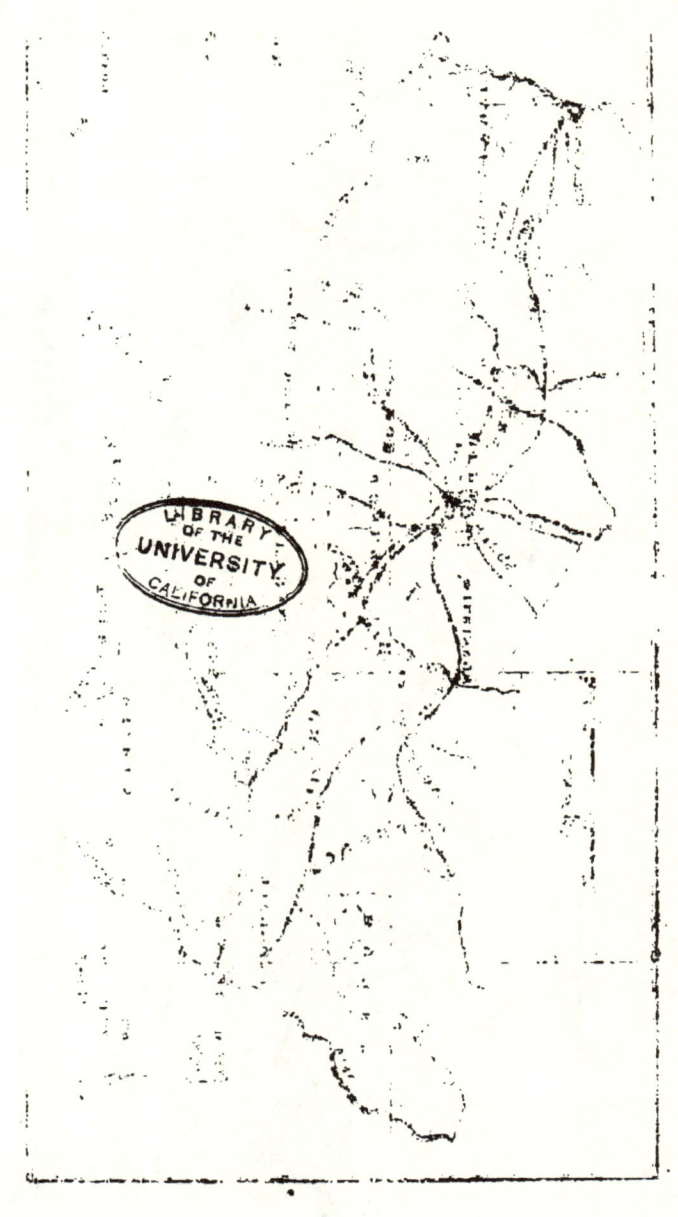

the Cretaceous strata, eastward, across the State, thus necessitating a modification of the geological map of Georgia, which has hitherto limited the Cretaceous to a strip of territory, traversing the central western part of the State. Second, the discovery of white kaolin, some of which ranks with the valuable South Carolina deposits, as "paper clay." Third, the experimental proof, that some of these kaolins, suitable for fire-clay, are more refractory, than any of the noted fire-clays of the United States.

GEOLOGY AND PHYSIOGRAPHY

East and South of the Appalachian mountains, which consist of folded Paleozoic strata, extends the so-called Piedmont Plateau, consisting mainly of ancient metamorphic and eruptive rocks. This "plateau" is a northeast and southwest trending belt, having an undulating or rolling surface with hill-tops lying, almost uniformly, in a theoretical plane, which dips away toward the sea. Bounding this area, on the seaward side, is the Coastal Plain, which differs widely from the Piedmont Plateau, in the nature of its rocks, in age, and in general aspect. The rocks here are sedimentary, and range from Lower Cretaceous to Pleistocene. The Metamorphic schists and gneisses of the Piedmont Plateau are crumpled and distorted; and the beds, if they may be called such, appear to stand edgewise. On the upturned edges of these rocks, are the almost horizontally lying, sedimentary strata, which make up the Coastal Plain. The boundary line, or the surface line of contact, between the Coastal Plain and the Piedmont Plateau is known as the Fall Line. Owing to the different geological conditions in the two regions, the rivers, traversing them, undergo a great transition in character along this Line. They cross the Piedmont Plateau, swift-flowing, often as cascades, and, still more frequently, as dashing rapids, the water tumbling over hard ledges and boulders of crystalline rocks. But, emerging from the gorges and comparatively deep valleys of the Piedmont belt, they begin a quiet, meandering journey, over the Coastal sediments, to the sea, the larger rivers becoming navigable at the Fall Line.

A study of the geological map of the United States will show, that, from New York southward, almost all the large cities are situated along this Fall Line, that is, along the Western margin of the Coastal Plain. In Georgia the important cities, so located, are, Augusta, Macon and Columbus, at the head-waters, respectively, of the Savannah, Ocmulgee and Chattahoochee rivers.

The topography, or surface features, of the Piedmont Plateau has resulted from forces diametrically opposite, in large measure, to those, which have outlined the surface features of the Coastal Plain. In the case of the former, the operative forces have been destructive; in the case of the latter, they have been largely constructive, and in part only, destructive. The present surface of the Piedmont Plateau has been attained by the removal of a vast thickness of overlying rocks, and is, we know not how much, lower than the original one. The rocks, of which it is composed, were once probably largely sedimentary, or water-deposited. They were traversed and overflowed by masses of igneous rock; and all have been so modified by the forces, which have acted on them, in the long interval of geologic time, since their formation, that their original nature is largely lost, and they have passed into the category, known to geologists as Metamorphic rocks. They are composed largely of quartz, feldspar and mica; and they are cut by innumerable quartz veins. These rocks, through the action of the atmosphere and surface waters, are constantly decomposed, and the rain and the streams are carrying the decomposed products to the sea. The difference between the rate of decomposition of the rocks and the rate of removal of the products, results in the surface zones, — soils at the top passing downward through less and less decomposed material, to the solid rock below, the latter being frequently stripped bare, and exposed in gullies and stream-beds. For a long time, even for a long geologic time, this process of lowering the surface of the crystalline rocks in the Piedmont belt, and the giving up of its material to the sea, took place, while the margin of the sea was far inland, with reference to its present position. The seashore may be considered to have been approximately along the line, which has been defined as the Fall Line. The products of the weathering and decomposition of the Piedmont crystallines, consisting of clay, sand and gravel, of varying coarseness, were emptied by the streams and rivers into the sea, and were distributed according to varying conditions. They finally built, along with the limestones, which were formed, the stratified beds of the Coastal Plain. These consisted, besides the limestone, of clays, marls, sands and gravels, the two latter having become, in places, sandstone and conglomerate. The materials are separated from each other, with varying degrees of perfection, and the sand grains and pebbles are fresh and regular, as they originated from the decomposition of the crystalline rocks; or they are rounded and smoothed, by the hardships of their journey to the sea, according to its length. After many oscillations, these sediments, emerged, finally, from the water; and the streams, which had formerly entered the sea at the Fall Line, sweeping across the formations, which their work had created, carved out the valleys, now existing in the Coastal Plain.

Owing to the comparative youth of the Coastal Plain, the features, which have been called constructive, that is, which resulted from the building up of the strata beneath the sea, are far from being obliterated. Therefore, it is seen, that, from the point of view, of the kind of rocks, their immediate origin, and the nature and origin of the surface features, the character of the Piedmont Plateau and the Coastal Plain are distinctly unlike. The latter was built, under the sea, of the materials, brought down by the streams from the former. The destruction of the one led to the growth of the other.

In the following story, the origin of the clays of the Fall Line is briefly told: —

Among the crystalline rocks of the Piedmont side, undisturbed, or *in situ*, are the clays, which have been gradually formed by the decomposition of the schists or gneisses. Among the sedimentary rocks, on the other side, are the beds of clay, which, originating in the crystalline rocks, were transported, with other materials; sorted out by the action of running water; and deposited in lagoons and quiet off-shore stretches. These were, in turn, buried by other clays, or beds of sand and gravel, sometimes having been first largely cut away, by rapid-flowing currents, resulting from changed conditions. The overlying beds would then rest on the very irregular surface of the clay, producing an "unconformity,"[1] a term used to describe the conditions, when the upper part of a bed, for example, has been removed (often irregularly), before the succeeding deposit is laid down. Sometimes, the same effect has been produced by the temporary emergence of the strata from the sea, their erosion by streams, their subsequent depression, and their burial by newly deposited strata, which rest on the eroded surface.

---

[1] See Plates IV, V and VI.

The location of the Fall Line in Georgia is shown on the accompanying map; and the relative positions of the Coastal Plain formations, with reference to each other, and to the rocks of the Piedmont Plateau, are shown in the diagrammatical cross-section.[1]

The decomposition of the rocks of the Piedmont Plateau now furnishes clays, *in situ*, and in low places, where the washings from the hills accumulate, which are suitable for many common uses. From them may be made a high grade of building-brick and a comparatively low grade of refractory ware. As a rule, however, they are too sandy and too full of various impurities; so, that their use is rarely more than for local purposes. On the other hand, the clays of the Coastal Plain region are very varied in character, often uniform in quality over wide areas, and of vast commercial importance. For the general reader, therefore, a brief outline of the nature and history of the Coastal Plain formations will be given.

THE COASTAL PLAIN

It should be understood, that the formations of the Coastal Plain region represent different geologic periods. They are all young, as compared with the rocks of the Piedmont Plateau, on which they rest. The history of the region is that of repeated subsidence and emergence; at one time, a land area being represented, dissected and eroded by streams, and, again, the whole, or a part, being covered by the sea, and reconstructed, or added to, by deposition on the old eroded surface.

Within the present area of Georgia, the Cretaceous, Eocene,

---

[1] See fig. 11.

Miocene and Pleistocene are represented by formations of varying thicknesses. A vertical cross-section would not show them with their original thickness, as deposited, on account of the erosion suffered, during the land intervals. Therefore, in large measure, just as they, collectively, are the "worked over" materials, formerly belonging to the crystalline rocks of the Piedmont Plateau, the successive formations are built up, in part, of the materials de-

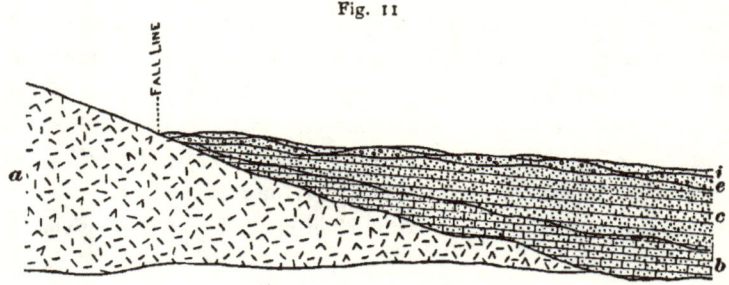

Fig. 11

Ideal Cross-Section, To Illustrate the Positions of the Various Formations along the Fall Line. a. Crystalline Rocks (Gneiss, Schist, Granite, Trap Etc.). b. Cretaceous Sands, Clays, Marls and Limestone. c. Tertiary Sands, Clays, Marls and Limestone. e. Lafayette Sand, Gravel and Clay. i. Columbia Sands.

rived from the underlying or older ones. The different periods are thus identified by "unconformities," and by fossils (when these are present), more readily, than by difference in the nature of the materials.

THE CRETACEOUS PERIOD

It was probably about the beginning of the Cretaceous period, that the subsidence took place, which brought the shore line to approximately the position of the present Fall Line, and which led to the earliest deposition of the formations, now constituting the

Coastal Plain region. Before the deposition of the great thickness of Tertiary strata upon the Cretaceous, much of the latter, especially along its land margin, was eroded, and re-deposited in the Tertiary seas. The subsequent removal of the Tertiary and overlying strata, has brought the Cretaceous rocks to the surface, over a considerable area in the western part of central South Georgia. This area is triangular in shape, with the base on the Chattahoochee river, the apex at Macon, and the northwestern side following, approximately, the Fall Line. Going down the Chattahoochee river from Columbus, a total thickness of Cretaceous strata, of about 1,640 feet, is seen, consisting of beds of clay, sand, gravel, marl and limestone. Along the river section, many interesting beds of clay exist, the consideration of which is left for a future report.

East of Macon, a narrow belt of the Cretaceous strata, probably the base of the series, has been traced by the writer, all the way across the State into South Carolina, some miles beyond the valley of the Savannah. The lowest group of the Cretaceous, formerly known in Georgia, Alabama and westward, as the Tuscaloosa, is now known as the Potomac group, being doubtless of the same age, as the strata of that group, which occurs further north, where the remarkably valuable clays of New Jersey are found, with the same relative position, with reference to the Fall Line, as in Georgia. The all-important clays of the Fall Line belt belong to this Potomac group.

THE TERTIARY ERA

The Tertiary Era is represented in South Georgia, by the Eocene and Miocene Periods, with a total thickness of nearly 2,000 feet. The Eocene formations have a thickness of nearly 1,600 feet, and consist of limestone, buhrstone, conglomerates and beds of sand, clay and marl, with occasional beds of lignite. These strata overlie the Cretaceous, and, in some places, completely over-reach this lower formation, and rest directly upon the Piedmont crystallines. The nature of the materials and their derivation have already been pointed out. During the Tertiary Era, there was at least one land epoch, with its resulting unconformity, and a great many minor oscillations, along the coast, changing the depth of the water, under which the rocks were deposited. The Miocene Period, with a total thickness, probably not exceeding 300 feet, rests upon the Eocene, and is apparently confined to the extreme southern and southeastern portions of the State, the depression, which led to its development, not having involved a great part of the Coastal Plain region.

THE COLUMBIA AND LAFAYETTE FORMATIONS

The formations of the Cretaceous and Tertiary succeed each other, without any long continued land intervals, geologically speaking; and, in spite of the unconformities mentioned, they rest, one on another, in a generally conformable series. At the close of the Miocene Period, however, South Georgia stood, for a long time, high above the sea, the great rivers carved broad and deep valleys, cutting down, in places, to a depth of 350 feet below the general surface level, for a breadth of several miles, the topographic features of the region becoming pronounced.

Following this land epoch, however, came a new subsidence, the shore-line moving inland to the present Fall Line, and, in places, somewhat beyond it. During this subsidence, when elevations, now 800 feet above the level of the sea, were washed by its waves, deposits of sand, clay and gravel were spread irregularly, some-

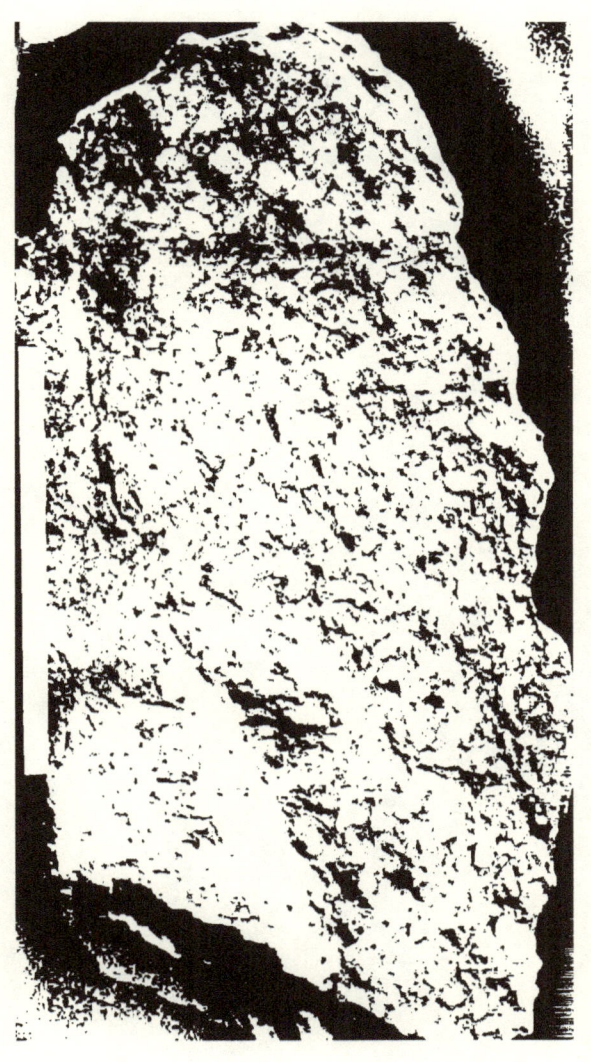

times to the depth of 150 feet, like a huge mantle, over the hills and valleys, which had been carved out, during the land epoch. The materials of this formation can be found all over southern Georgia. They rest on the hill-tops, on their slopes, and on the floors of the valleys, and greatly interfere with geological research. This formation is known as the Lafayette.

After a comparatively brief subsidence, there was a re-emergence, when new valleys were cut, often through the Lafayette, and penetrating the underlying strata. Then took place a final submergence and a deposition of the superficial sands and gravels, which make up the Columbia formation. The Columbia and Lafayette formations are unconformable, with reference to each other; they also lie unconformably upon the Tertiary and Cretaceous strata; and, along the margin of the Coastal Plain, even upon the crystallines of the Piedmont Plateau.

GENERAL ASPECT OF THE FALL LINE BELT

Journeying from Columbus to Augusta, *via* Macon, by rail, or over the dirt roads, the Fall Line will be crossed and re-crossed, and the observer will notice, now the features of the Piedmont Plateau, and, again, those of the Coastal Plain. On the Piedmont side, he sees the well-rounded hills and deep red soils, which characterize its ancient formations. Occasionally, and especially in the vicinity of streams, are seen bare stretches of crystalline rocks; while, here and there, exposed in the gullies and road-way cuts, are to be seen the various transitions between the fresh crystalline rocks and the products of its decomposition, which constitute the soils. The schists are brilliantly colored, especially where decomposed, — red, yellow and greenish shades predominating. The soil is comparatively fertile, and yields abundant crops.

On the Coastal Plain side, where the highest elevation is about 750 feet, the well defined hills are replaced by broad, flat-topped plateaus and wide reaching terraces, the latter, particularly, predominating in the vicinity of the great rivers. Occasionally, streams and railroad cuts penetrate through the sedimentary strata and the crystalline rocks, below, it being often hard to distinguish, where the one ends and the other begins, the lowest of the sediments being composed of the coarser elements of the underlying crystallines, large boulders of the latter frequently occurring. The contact of the Cretaceous with the crystallines is best seen, on the Alabama side of the river at Columbus.[1]

The Lafayette and Columbia sands and gravel, which cover the Coastal Plain, may be seen, along much of the whole distance, to have smoothed out the minor irregularities of the old land surface. They make a less fertile soil, as a rule, than that found on the Piedmont side of the line; and these formations are responsible, for most of the barren sand-plains, found in South Georgia.

The Lafayette gravels are characterized by cross-bedding, rapid transition from coarse to fine material, and generally, by a deep-red color, resulting from the oxidation of iron-bearing materials. It frequently has a mottled appearance, which is noticeable in the sides of gullies and cuts. The prevailing red color is frequently broken, by a net-work of gray and blue tints, which mark the former position of the roots of trees and shrubs, where the iron oxide has been "reduced," by the organic compounds resulting from their decay.

The Cretaceous beds, that is, those of the Potomac Group, which

---

[1] See Plates XV, XVI and XVII.

occur along the Fall Line, are visible at many points in cuts along the Central Railway, particularly between Columbus and Macon, and for many miles east of Macon, both on the Central and Georgia Railroads. They consist mainly of white clays and sands, sometimes wholly differentiated from each other, the clay being remarkably free from grit or sand, and the sand beds, in turn, being entirely free from impurities of any kind, ready, indeed, to be manufactured into the best plate-glass.

The Tertiary beds are exposed, at many points along the dirt roads, chiefly east of Macon, excellent sections of fossiliferous marls, of Eocene age, being exposed in the railroad cut at Summit, in the southern part of Jones county. One of the most interesting sections along the Fall Line belt is at Rich Hill, which has an elevation of 750 feet. It is capped by Lafayette materials, and exposes in deep gullies on the southern side, one hundred and fifty feet of Tertiary strata, resting uncomformably upon the white Cretaceous sands and clays at its base.

NOTES BY LOCALITIES

COLUMBUS

Columbus is situated at the Fall Line on the Chattahoochee river, down the course of which a magnificent section of the Coastal Plain formations is exposed. At Columbus itself, however, there are to be seen only the crystalline rocks of the Piedmont belt, and overlying them, unconformably of course, the Potomac or base of the Cretaceous series, on the eroded surface of which rest the clays and gravels of the Lafayette and Columbia formations, the latter lying unconformably on the former, so that three distinct unconformities are recognizable.

The crystallines are gneissic and schistose in character, and are beautifully exposed at low-water in the river bed, above the steamboat landing. The best sections are exposed on the Alabama side, where a small stream has dissected the plain, and cut down into the underlying crystallines. A general view of this locality is represented in Plate XVI, which shows some of the general geological and topographic features of this region, and is a beautiful illustration of stream-erosion. The Potomac formation is clearly seen to rest, as shown in Plate XVII, on the old eroded surface of the crystallines, which are very much decomposed, although the structure of the rocks still remains. Between high- and low-water marks in the river-bed, where the overlying formations have been stripped off, the transition from the wholly decomposed gneiss, down to the fresh, unaltered rock, is strikingly apparent. The decomposition of these rocks has given to the gently sloping bed of the river, as seen at low-water, an exquisite coloring effect. The most delicate shades of green, yellow and pink are harmoniously blended, and set off by the deep orange and reddish tints of the precipitous banks, which rise to a height of twenty-five or thirty feet, in the back-ground. A general view of these decomposed and colored rocks is shown in Plate XV; while, in Plate XVII, is illustrated a near view, showing the distinctness, with which the banded structure is retained, even where the gneiss is practically wholly decomposed.

The Potomac formation consists here of Arkose, or the materials of the decomposed gneiss, but slightly sorted, so that it is sometimes difficult to determine where the gneiss ends and the Potomac begins. It furnishes however, in places, beds of clay, derived from the decomposed crystals of feldspar of the gneiss, which, after washing to remove the sand, have been used, locally, for the manufacture of pottery. The Lafayette formation, with its characteristic color and sand and gravel beds, is exposed in cuts and ravines, on the higher ground, especially on the Alabama side. The local surface features are here determined, in a general way, by the formation last laid down, that is, by the Columbia. This constitutes the plain or terrace, on which the city of Columbus stands, and is made up of variegated sand, alluvial clay and loam. The general section seen on the Alabama side of the river is, descending from the surface, first, red and orange loam, then mottled purple and reddish clays resting on cross-bedded, gray, white and pinkish sands, somewhat consolidated, which, in turn, rest upon the varicolored decomposed gneisses.

CLAY INDUSTRIES AT COLUMBUS

The manufacture of clay products at Columbus is confined to the making of bricks and common pottery. The total annual product of brick is about 10,000,000, which are worth, on an average, $5.00 per thousand. The clay used is the alluvial material of the Columbia formation, and is quite plastic and free from grit. It is white and bluish colored, and varies in thickness from ten to sixteen feet.

The following plants were in operation, at the time of the writer's visit to this locality: —

SHEPHERD BROTHERS' WORKS

The works of this firm are situated on the plain of Columbus, just south of the city limits. The product is chiefly stock-brick; but some face-brick are manufactured, by re-pressing. The stiff-mud process is employed, the machine used being a "side-cut" Dalonian.

C. W. RAYMOND'S WORKS

Mr. Raymond has a plant similar to that of Shepherd Brothers, situated three or four hundred yards from the Georgia and Alabama Railroad, with a capacity of 80,000 bricks per day.

G. O. BERRY'S WORKS

Mr. Berry's plant is at 10th Avenue and 8th St., where about 35,000 bricks per day are produced, on an average of twenty days in the month. The bricks are made in a Chamber's end-cut machine. "Temporary" kilns are used at this, as at the other Columbus yards. Mr. Berry manufactures, also, an excellent quality of jugs, which are burned in a small rectangular kiln.

J. L. MATHEWS' POTTERY

Mr. Mathews operates a pottery, making jugs, churns and flower pots, in a small way, burning in a small, rectangular kiln. The clay used comes from the brickyard of Messrs. Shepherd Brothers.

THE BUTLER DISTRICT

Going east from Columbus, the mottled clays and cross-bedded sands of the Potomac formation may be traced all the way to Macon, across the broad divide between the Chattahoochee and the Flint rivers, to Everett Station; and, again, from there over into the valley of the Ocmulgee. Overlying the Potomac, there may be seen, at many points, from ten to thirty feet of reddish gravel and sands, with occasional streaks of white clay, belonging to the Lafayette. The Potomac sands are strikingly cross-bedded, and contain lenses of very coarse pebbles. The beds, as a whole, have little continuity, thickening rapidly, and as rapidly disappearing, to be replaced by other beds. The clay is frequently sandy. The overlying sand and gravel is not always red and orange, but, occasionally, white and gray; and it is sometimes indurated, and stands out in massive walls. In many places, the incoherent, fine-grained, buff-colored sands, varying in thickness from two to twelve feet, appear to lie unconformably on the Lafayette; but, sometimes, they apparently grade uniformly into it. They may belong to the Columbia formation. The Potomac formation, although generally mottled with bright-red and purple colors, is frequently white, and, again, drab and bluish. It has a wavy, irregular surface, on which the overburden rests, with distinct unconformity. This formation, itself, rests, like a mantle, on the underlying crystallines, just as the Lafayette and Columbia formations mantle over the hills and valleys of the Tertiary and Cretaceous. In going east, it rises from the valley floor of the Chattahoochee at Columbus (230 feet above the sea), to an elevation of 670 feet, at Butler. On a railroad trip between Columbus and Fort Valley, the clays of this formation are frequently seen to rest directly on the upturned beds of gneiss, which are weathered to a depth of several feet, and which resemble closely the clay itself; but, when crossed by streams, the loose material is cut away, and the solid gneiss is exposed.

At Butler, a number of openings have been made in the Potomac clays, and some of the best properties have been purchased by a syndicate, with the object of excavating and washing these clays, for the wall-paper industry. One of the Butler localities, where developments have actually been made, is two and one-half miles west of the railroad station, where the white and mottled clays of the Potomac formation are exposed at, or near, the top of the circumference of the broad plateau. Where the clay is exposed, there is practically no overburden, for many yards along a broad belt, running around the plateau; but, beyond this belt, a gentle slope is formed of Lafayette gravel, which reaches an approximate thickness of some twenty feet. The clay here has a thickness of twenty to twenty-five feet, and is underlaid by sand. In places, it is snowy-white, when dry, and cream- and drab-colored, when moist. At other points, it is orange-colored, and stained with iron along the joints, and, sometimes, in the clay itself. It is largely free from grit; but it always contains some, which is often very coarse.[1] The clay itself is sometimes coarse-grained, and contains scales of mica. The clay is massive and jointed. At the time of the writer's visit, a company had erected a small drying-shed, and some experimental apparatus for grinding and washing clay. A few car-loads of the product had already been shipped to northern markets. Across

---

[1] The frequent occurrence, in this formation, of the very coarse grains of sand, scattered, in abundance, irregularly through the beds of very fine-grained clay, presents a problem difficult to explain, unless it may be, that the coarse grains were floated on the surface of the water to places above their present position, by the action of "surface tension"; while the fine-grained clay was still settling in the very quiet waters, which would not transport the sand, once it were completely immersed.

KILN AT A SMALL COUNTRY POTTERY, CRAWFORD COUNTY, GEORGIA.

the valley, perhaps a mile in width, the clay occurs, in much the same manner. It is, however, very dark-colored, dense and more plastic, and has less grit, often none at all. When first mined, it is chocolate-colored, and, in places, drab. Upon drying, though still dark, it becomes much lighter in color. It makes a good pottery clay, and will surely find a market, in time. There are many other localities in the vicinity of Butler, where these clays may be found. Among these, is one, on the property of Mr. J. J. McCants, about four miles west of the railroad station.

The surface of the country about Butler is generally sandy, being covered usually with a loose, buff sand, which rests on more or less compacted sands, sometimes replaced by a hard, waxy, but sandy, clay.

It is often difficult to determine the geological formation, with which one is dealing, in this vicinity, owing to the similarity in the deposits of the different groups, particularly those of a sandy nature, which, being alike in origin, resemble each other very closely, and all the more so, when affected by surface conditions, oxidation, de-oxidation, disturbances by the roots of plants and trees, and by the slides and washings down slopes.

THE RICH HILL AND CRAWFORD COUNTY LOCALITIES

Going northward from Fort Valley into Crawford county, a broad plateau, about 525 feet above the sea, is traversed for a number of miles, with scarcely any perceptible change in elevation. A gradual rise then takes place, until an elevation of about twenty feet above the level of Fort Valley is reached, where a very rough, broken country is encountered, six or seven miles north of that place, with hills, which ascend from 80 to 90 feet above the highest part of the Fort Valley plateau. This broken country continues, until Rich Hill is reached, a short distance southeast of Knoxville, which rises abruptly 165 feet above the surrounding hills, or to an approximate elevation of 670 feet above the sea.

The surface of the Fort Valley plateau consists of a red, waxy sand and clay formation, cut into gullies by running water, as though with a chisel, and full of *pot-holes*. In a few places, fine-grained plastic-clay seams are encountered; also thin layers of coarse pebbles.

Entering the broken country north of this plateau, near the base of the steep hill, a clay, probably the Potomac, is encountered, in places, white, but more generally mottled with red, purple and yellow tints. The surface of this clay is found at different elevations at different points; and it is impossible to say, whether this is due to extensive erosion, preceding the deposition of the overlying beds; to flexures; or to its rising and falling with the contour of the old Piedmont topography, which underlies it. The greatest thickness of this clay, seen at any one point, was thirty-five feet. Overlying it is a red and yellow, tough, sandy clay, with numerous plastic-clay partings, with the upper part more sandy, and strikingly cross-bedded. The slope above this is covered with a white and gray sand, on the south side of Rich Hill, which is traversed by tremendous gullies. The following section is seen here, in descending series: —

    1 Tough, red clay, overlying sands of various colors, and containing at the base more or less lignite.. 45 feet
    2 Gray and drab, fine-grained clay, which breaks concentrically on drying. When cut with a

knife, it gives no evidence of grit, but shows a smooth, polished surface. On drying, it becomes hard and porous, and is almost white. It resembles the clay of Fitzpatrick, described on another page .... 30 feet
3   Limestone beds, composed almost entirely of shells and shell fragments................... 25 "
4   Red and buff sands .... 15 "
5   Cross-bedded sands and clay, unconformably overlaid by No. 4.......................... 50 "

There is an apparent unconformity between Nos. 1 and 2. Some of the beds are highly fossiliferous, containing crinoid and bryozoan remains, sharks' teeth and numerous bones of vertebrates. The surface of some slopes consists wholly of a loose sand, of shell fragments, largely delicate bryozoan remains. The lower thirty feet of this section consists of fine, white sand and clay. Two or three feet from the base, there is apparently another unconformity. These white sands and clays are probably Potomac. They vary rapidly in character, and replace one another, in a manner indicating frequent changes from strong currents to quiet waters.

Some of the colored beds in this section are curiously "whitewashed," with a thin layer of fine, white clay, which the waters have brought down from overlying seams. In places, the down-dripping water, holding in suspension particles of white clay, is evaporated; and stalactites, sometimes in the shape of broad, wavy sheets, suspended from overhanging strata, are found, just as limestone stalactites are produced, by the evaporation of water carrying the burden, except that in this case, the clay particles are held in suspension, instead of in solution.

A general view of the Rich Hill section is shown on Plate III. At many points in Crawford county, the clay, which is doubtless

Potomac, is used for the manufacture of common pottery, an industry carried on in this country in a primitive way, and under conditions, which make its continuation remarkable. There are certain families, which are brought up, generation after generation, as potters, the business being carried on chiefly at odd times in the winter, and, at intervals, in the summer during the making of crops. A small plant, typical of this country, is illustrated in Plates IX and X; it is located at Dent, a post-office in the interior of the county. Plate IX shows a primitive kiln, which, however, answers its purpose quite satisfactorily; and Plate X, the clay mixer. The products of this pottery are sold mostly by peddling, the articles being carried in wagons, so the writer was informed, sometimes to a distance of 75 miles, and disposed of, partly for cash and partly in trade. These potteries are located mostly in the vicinity of Dent, Pine Level and Gaillard's. Among the proprietors of different establishments, are Messrs. J. Becham, W. Becham, S. Becham, J. S. Newberry, J. N. Merritt, Thos. Dickson, H. N. Long and H. D. Marshall.

The kilns, like that in the illustration, hold from 300 to 500 "gallons" per burning. They are rectangular, and have an intermediate draft. The clay used is a mixture of "chalk" (white Potomac clay), with mud obtained from the swamps. The clay is hauled to the potteries in wagons, over several miles of sandy roads. For glazing, these potters use a mixture of wood ashes, lime and sand. This is colored by the addition of powdered limonite. It makes a fairly good glaze; but it is rough, owing to the coarse sand contained.

THE MACON DISTRICT

Macon, the chief city of Middle Georgia, is located in Bibb county, at the Fall Line, the head of navigation on the Ocmulgee river. The business portion of the city is built on a broad terrace, the second one above the river. The suburbs and much of the southern portion of the city is on the hills, flanking the western side of the valley.

The valley of the Ocmulgee, here, is broad, with gently sloping sides, on which three terraces are usually found. The geological formations and their relationships differ from those about Columbus, in that the Cretaceous groups are absent, with the exception of the Potomac, which itself is wanting in the immediate vicinity of Macon, having evidently been entirely removed in the valley of the Ocmulgee, before the deposition of the Lafayette and Columbia formations.

The topography of the country about Macon is typical of that of the Fall Line in general. North of the city, are the red hills of the crystallines; south of it, the terraces of the Ocmulgee, and the broad sand and gravel-capped plateaus, — *divides* between minor streams, which flow into the great river. These are not confined to the valleys, but sweep about over the whole country, south of the Fall Line, and broaden into the sandy plains, called also, technically, terraces. The Fall Line crosses the Ocmulgee river, beneath the Macon foot-bridge; and here are exposed boulders and out-cropping ledges of the hard Piedmont crystallines. Overlying them, are the Columbia terraces; and, east and west, mounting over, but capping, the Tertiary heights, are the white, red and orange gravels and sands of the Lafayette.

Beyond the limits of the Ocmulgee valley, on both sides of the river, railroad and highway cuts expose the Potomac clay and sands, which vary from the occurrences, as described at Columbus, Butler and the Rich Hill district, in that they are less highly-colored, and that the remarkable, mottled appearance has grown less and less marked coming eastward. The characterstic cross-bedding is retained, and white sands and gravels, which may belong to the Cretaceous, or which may, however, be as recent as the Lafayette, overlie these clays, unconformably, occupying deep channels and wide, irregular areas in them, and containing much material, derived from their beds, and mixed in with the sand and gravel. Sometimes, this is present, as partings; and, sometimes, large fragments and even boulders of the white clay occur. Characteristic views, showing this phenomenon, may be seen in Plates IV, VI and VII. They will be described more minutely on another page.

One of the best sections of the Lafayette, in the vicinity of Macon, is on the Forsyth road in Vineville. Here, are cross-bedded gravels, containing large quartz fragments, and some mica, of a pink and red color, cut out by cross-beds of a very coarse gravel, with pebbles two to four inches in diameter, the whole coated with red iron oxide, which gives the mass its characteristic color.

North of Macon, *district 13, lot 338, N. E.*, there are outcrops of the decomposed crystalline rock, unlike the gneisses in the vicinity of the Macon bridge, and showing rather the characteristics of schist.

On the estate of Mr. R. E. Park, near Holton, there is a cross-bedded, gray, sandy clay, associated with a friable buff and pink sandstone, loosely consolidated, and containing many large pebbles. This clay seems to have been derived from the decomposed schists in the vicinity, the products of which have been washed into low-lying ground. It may, however, be a remnant of some of the Coastal Plain formations, perhaps the Lafayette.

About seven miles southwest of Macon, on the Central Railway, at a place called Rutley, there is a good exposure of the Potomac. At the surface, here, there is a reddish and buff compact sand, six feet in thickness, carrying a two-inch seam of hard ferruginous sandstone. Beneath this, is a cross-bedded gravel, six feet in thickness, and composed of coarse pebbles. These are probably Lafayette; they rest unconformably on six or eight feet of white, bluish and pinkish Potomac clays, known locally (as is the case everywhere along the Fall Line), as "chalk." The clay is very irregular in quality, being sandy and somewhat iron-stained; but, over considerable areas, it is pure white and quite free from particles of sand.

Entering Macon on the Houston road, near the foot of a long hill, down the slope of which Lafayette gravels are exposed, is another section of the Potomac clay, or "chalk."

On the Jeffersonville road, east of the river, a number of sections occur. Here, in the Lafayette formations, are found cross-bedded sands and gravels, containing large, angular, transparent quartz pebbles. These sands are much mixed with white kaolins, which occur in places, as thin seams. The formation mounts the hills, and flanks their sides. In the railroad cut near this road, is a stratum of cross-bedded sandstone, overlying the pure white clay or kaolin, the surface of which is completely coated with a wash of red clay, coming from the overlying formation. The mottled appearance of the red clays, due to de-oxidation by organic matter from the roots of trees, is here apparent.

At a few points along the Jeffersonville road, fossiliferous strata of Tertiary age are exposed in gullies, by the way-side.

Six miles northeast of Macon, on the Griswoldville road, where the highway is crossed by the railroad, there is exposed a section, forty feet in thickness, the strata all showing evidence of extreme changes in the rapidity of currents, cross-bedding being prominent, and the sudden transition in the size and nature of the materials being a striking feature. At the base of this section, are ten to twenty feet of white cross-bedded sands, with lenses of kaolin and mica, both of which are also largely mixed with the sand. This is doubtless Potomac. Unconformably overlying this, are ten to twenty feet of red and mottled micaceous clays and sands. Over these, but often wholly wanting, are two to four feet of gray and buff sand and loam. Similar sections are to be seen at many other points.

The clay industry of Macon is confined to the manufacture of building-brick and sewer-pipe. The clay used is chiefly that of the Columbia formation, which is taken from the lower terraces near the river. But some clay is brought by rail from points in Baldwin county.

THE H. STEVENS SONS CO.'S PLANT

The largest clay manufacturing company is The H. Stevens Sons Company, which produces sewer-pipes and drain-tiles, of all varieties. Their works are located just south of the city limits. They have a large building, 60 x 360 feet, with two stories and a basement; also 10,000 feet of shed space. The machinery includes the usual variety of disintegraters, pug-mills, wet-pans, pipe-press, elevators etc. The wares are burned, in six circular down-draft kilns. The present capacity of the plant is 500 car-loads per year. It was established in 1887, and is the leading sewer-pipe plant of the south.

THE CLAYS OF GEORGIA

PLATE X

MIXER AT A SMALL COUNTRY POTTERY, CRAWFORD COUNTY, GEORGIA.

The following firms are engaged in manufacturing building brick: — A. C. Earnest & Co., Peter Harris, The Macon Brick Works, C. J. Tool, Geo. Anderson, C. C. Stratton and G. J. Blake. These plants are all located on the lower terrace, and near the river.

GRISWOLDVILLE

Griswoldville is centrally located, in an area of the best Potomac clays found along the Fall Line. This area includes the southern half of Jones and Baldwin, and the northern part of Twiggs and Wilkinson counties. As a whole, it is a part of the water-shed between the Oconee and the Ocmulgee rivers.

The town of Griswoldville is situated on the main line of the Central Railway, about ten miles east of Macon, and five miles south of the Potomac belt. Within the area mentioned, there are many geological sections of the formations, which have already been described, and a large variety of clays, suitable for commercial purposes. These are found among the decomposed crystallines; in the Potomac formations; at many points among the Tertiary strata; distributed over a wide area in the Lafayette; and in the Columbia formations in the valleys of the Oconee and Ocmulgee. The Columbia clays are utilized at Macon, in the latter valley, and at Milledgeville, in the former. The Potomac clays, found in this area, are of the best quality; and they occur in the greatest abundance.

J. R. VAN BUREN & CO.'S WORKS

At Griswoldville, the Potomac clays are found at a number of points on the property of Messrs. J. R. VanBuren & Co. They are exposed in railroad-cuts, and in gullies, which occur along the terrace followed by the Central Railway. This terrace is about 100 feet above the Macon bridge. The surrounding hills, composed of Tertiary strata, mantled with Lafayette, rise to an average height of 160 feet above the plain, or 260 feet above the terrace, on which the business portion of the city of Macon stands.

One mile west of the VanBuren residence, in a railroad-cut two or three hundred feet long, the Potomac clay is characteristically exposed, overlaid by Lafayette gravels. An illustration of this section is shown in Plate II. At the base may be seen the snowy-white kaolin and the white cross-bedded sands of the Potomac formation, above which are the red and occasionally white gravels and clays of the Lafayette. Near the contact of the two, the Lafayette consists largely of materials taken from the Potomac.

On the south side of the railroad, the white clay, which is sandy in patches, has a thickness of from ten to twelve feet, while the overlying red and white cross-bedded Lafayette has a thickness of fifteen or twenty feet. On the north side, the surface of the Potomac clay is lower, and, in places, wholly cut out by Lafayette gravels. South of this cut, the Potomac clays are exposed in old fields over wide areas.

Two miles beyond the first railroad cut, and still on the Van-Buren property, is another railroad-cut, where again the Potomac clay is seen, under similar conditions. From these two railroad-cuts, a large amount of clay has been shipped to Chattanooga, Tennessee, to be manufactured into fire-brick and refractory wares.

East of the VanBuren residence, from one to two miles, cuts along the railroad still expose valuable beds of the Potomac clays.

The clay on this estate is, in many places, entirely free from grit, and, therefore, suitable for the manufacture of wall-paper, for which purpose it is specially adapted. Some of it, however, is more or less colored, usually, a drab or cream color; and occasionally it is sandy, or contains fine grit. This, however, can be removed by washing, there being an abundance of water in the vicinity, thus making the clay suitable for paper manufacture. An average sample of clay, taken from the first railroad-cut, varies in color, from snowy-white to a creamy tint; and it is about as hard as chalk. It is fine-grained and tough, and has a soapy feel. An average sample of the clay from the field east of the VanBuren residence, and south of the railroad-cut, is cream-colored, when damp, and almost white, when dry. It is very fine-grained. Muscovite scales are not seen with the naked eye, and but very little of this mineral is detected, by the use of the lens. It absorbs water and disintegrates rapidly, and becomes somewhat plastic; and it shows a trace of grit to the feel. An occasional grain of sand may be seen under the microscope, with a high power. The only impurity observed is muscovite, although there is less of this mineral, than in most other Potomac clays of this region. Quartz, as a rule, is wholly absent; and, on the whole, it is one of the purest clays studied. Some specimens are locally stained with iron oxide, apparently due to the alteration of ferruginous minerals; and, in these, the lens reveals scattered muscovite plates and some magnetic crystals, but no quartz. Under the microscope, a multitude of muscovite plates and prisms are visible; also a trace of magnetite, but still no quartz.

A chemical analysis of the pure variety gave the following result: —

Hygroscopic Moisture	0.57
Combined Water, Carbonic Di-oxide Etc	13.08
Combined Silica	44.94
Alumina	39.13
Free Silica, or Sand	1.23
Ferric Oxide	0.45
Lime	0.18
Magnesia	0.11
Potash	0.51
Soda	0.63
Total	100.26
Clay-base Present	97.34
Fluxing Impurities	1.88

Experiments, as described on pages 45 and 46, show, that the uncompressed dried clay from this locality absorbs 112 per cent. of its weight of water. On increasing the density of the clay by 25 per cent., it absorbs 114 per cent. On increasing it another 25 per cent., or 50 per cent. in all, it absorbs 109 per cent. of its weight; showing, that the density of the clay powder does not materially affect its absorptive power, so long as it is free to expand in at least one direction.

This clay, when wet and moulded into briquettes, shrinks, on drying, but .8 of one per cent. Its tensile strength, after drying, is 25 pounds per square inch. In refractoriness, it is almost equal to the highest grade Seger cone manufactured. That is, it will stand more heat, without fusing, than any clay in the United States, studied, except the similar clays in neighboring localities. Its specific gravity is 1.76.

Experiments made on a sample, collected from a field adjoining this locality, show, that it absorbs 85 per cent. of its own weight of water, and shrinks, on drying, about one per cent. of the length of a briquette. The maximum tensile strength is 41 pounds, per square inch. Its specific gravity varies from 1.62 to 1.80. The burned product is white to fine cream-colored, according to the conditions of burning; and, unless great care is used, it crackles and bursts open, — the only objection to its use for the manufacture of china ware. It is exceedingly refractory, and fuses with Seger cone 35.

An average sample of the clay from the railroad cut, on the VanBuren property, east of the residence, has the following characteristics: —

Some of the hand specimens show blotches and seams, of a reddish-brown mineral, which, under the lens, seems to be hematite, or possibly turgite. Under the microscope, grains of quartz are visible, varying from .001 to .004 of an inch in diameter, although some are recognizable, as small as .005 of an inch. There are some few muscovite scales; and scales of kaolin are grouped together, in more or less coherent aggregates.

An average sample from the second railroad cut, west of the house, shows a massive clay, with numerous muscovite scales and small crystals of magnetite. This clay readily disintegrates in water, and is moderately plastic. Under the microscope, with a high power, many muscovite prisms are observable; and scales, averaging in size about .001 of an inch in diameter. This clay shrinks, on drying, 8 per cent., and has a tensile strength, dried, of 36 pounds per square inch. Its fusing point is between those of Seger cones 34 and 35. It burns to a creamy color, and crackles, when burned by itself.

THE PROPERTY OF MRS. SALLY JAMES

Northeast of the estate of Messrs. VanBuren & Co., are numerous other exposures of a high-grade clay (belonging to the Potomac formation) on the estate of Mrs. Sally James. This clay, or "chalk," is here exposed in gullies, and is seen, also, in the vicinity of the estate of Mr. G. W. James, where there are small outcrops. One of these is just east of the James residence, where the following section may be seen: —

1	Red Lafayette gravel	10 feet
2	White Potomac clay, carrying a thin, half-inch seam of iron ore	3 "
3	Micaceous sand	2 "

The clay is of good quality. At the point, where it is exposed, it has been largely cut out by the overlying gravel; but there are doubtless thicker beds, at other points in the neighborhood, now buried by the overlying gravels. It is probable, also, that other similar clays lie beneath the bed of micaceous sand, seen at the base of the section.

The second exposure shows from four to six feet of white clay, unconformably overlaid by 15 feet of red and white cross-bedded sands, which contain lumps of white clay, derived from the underlying stratum.

In the railroad-cut, near the James residence, is an exposure of red sand and loam, containing nodules of limestone, which overlie the white, micaceous sands, and contain, also, numerous scattered balls of white kaolin. In all probability, a white Potomac clay lies beneath this formation.

LEWISTON

At Lewiston, a station on the Central Railway, a few miles east of Griswoldville, in the extreme northwestern corner of Wilkinson county, thick beds of a remarkably pure and snow-white variety of clay occur.

THE LEWISTON CLAY WORKS

This clay is mined by The Lewiston Clay Works, established in 1893. Mr. J. W. Huckobee is the Superintendent of the works. The product is shipped north, the highest grades being consumed by the wall-paper industry, and other grades, in the manufacture of encaustic tiling and similar wares. Owing to the removal of the clay, by the company, from its pits, remarkably good sections, are exposed. There are at present three of these openings or pits, situated on the north and northeast slope of the hill, and but a few yards from the railroad track.

The western pit has been abandoned, owing to the thinning out of the clay bed. The clay and fifteen feet of gravel stripping have been removed, here, over an area of a quarter of an acre. The stripping is carried away in hand-barrows, and dumped on the slope of the hill, below the outcropping beds of the clay strata. The clay itself is worked by pick and shovel. It all requires careful sorting, and is separated into different grades, much of it being discarded, which, in time, will be valuable for markets, now barred by high cost of transportation. Only that clay, entirely free from grit, is utilized; and this is sorted on the basis of color, there being three grades: — First, the pure white; second, the yellowish; and third, the spotted. The last mentioned is only faintly mottled, the general effect being white, exceptional specimens, only, showing grayish and pale-yellow tints. When dry, it is whiter, than when first mined. Even that, which has a distinct yellow tint, when fresh, dries very nearly pure white. After mining, the clay is placed in open drying-sheds, of which there are four,[1] the lumps of clay resting on shelves, made of bars two to three inches apart, and arranged to permit free circulation of air. After drying, the clay is broken up and rammed down, by use of a mole, into hogsheads, in which it is shipped. These hogsheads hold, each, about one ton of the dry clay. They are made on the place, at a cost of about one dollar apiece. The plant is small and inexpensive, consisting only of drying-sheds, an office, a small store-house for tools, open sheds, where the casks are manufactured, and a short, wood-rail tram and car.

OCCURRENCE OF THE CLAY

The clay varies from three to eight feet in thickness, and is of massive structure. Its upper surface is very uneven, and there is a marked unconformity, between it and the overlying sediments. This is illustrated in Plate IV, where the workman in the foreground is pointing to the lines of unconformity, he himself standing against a wall of hard white clay. This view shows, also, in the overlying strata, the occurrence of balls of white clay, derived from the lower bed and torn loose, probably by the action of the coarse pebbles upon this bed. They cannot have been transported far, or they would have been disintegrated, and deposited in another manner, in more remote and quiet waters.

---

[1] See Plate XI.

THE DRYING SHED AND A CLAY-PIT OF THE LEWISTON CLAY WORKS CO., IN JONES COUNTY, GEORGIA.

Overlying the clay proper, is a stratum of eight feet of loose, unconsolidated sand, sometimes cross-bedded. Then occurs a lenticular-shaped bed of clay, with a maximum thickness of two feet, six inches; then red clay and sand, which continues to the surface, with a maximum thickness of six feet.

In the east pit, the following section is seen, from above downward: —

1. Red and yellow, clayey sand, with seams of laminated clay; also thin seams of brown iron ore, containing many coarse pebbles .... 6 feet
2. Irregular siliceous beds, resembling quartzite, and containing drusy quartz cavities and many fragments of shells .... 4 "
3. White sand, free from iron stain, cross-bedded in places, containing mica and kaolin, and also nodules or fragments of white clay, in the upper surfaces of which are sharply-outlined, pear-shaped cavities, each filled with a yellowish clay. These cavities vary in diameter from one-fourth to one inch .... 7 "
4. White, gritless clay, or kaolin .... 2 "
5. White sand .... (?)

The layers of sand and laminated clay of this locality show signs of crumpling, due to lateral compression. Similar evidence of disturbances is seen at many other points, where slight crumpling or folding, faulting on a small scale, and the presence of numerous slickensided surfaces in beds of clay, show the occurrence of a crushing force, at some comparatively recent time.

The interesting features, which are shown to best advantage at this locality, are the nature of materials; evidences of crumpling and lateral disturbances; a distinct unconformability; and the presence of a large amount of kaolin, resulting from the decomposition of feldspar fragments, which once constituted much of the mass of sand of this region (that is, it belongs to the class of "Metasedimentary," as given in the classification of clays, on page 10). The presence of fossils in the materials, particularly the clay here, is also very interesting. The true horizon of the beds, carrying them, has not yet been determined. They may be Tertiary or Upper Cretaceous. It is probable, however, that they are Lafayette, the forms having been transported there from the beds, where they originated, in the Tertiary rocks, of which the higher lands are composed.

As a rule, the gravels are of a sub-angular character, and are but little water-worn. In fact, the minerals of this whole formation seem not to have traveled far, but to have come from the gneissic and granitic belt immediately north, and to have been deposited by currents, changing rapidly in direction and rate of flow. It is worthy of note, also, that the flat, sandy plain, in which the clay formation occurs, has the same elevation, namely 260 feet above the Macon bridge, as have the hills about Griswoldville.

An average sample of the clay, from the Lewiston Clay Works, has a faint-yellow tint, which becomes practically white, on drying. It is massive and somewhat friable; and, so far as can be seen with a hand lens, it is nearly free from impurities. It absorbs water rapidly, dried clay taking up 108 per cent. of its own weight. Examined under a microscope, with a high power, but few impurities are seen. These consist of muscovite prisms, occasional grains of magnetite, and some few quartz grains, the largest being .05 of an inch in diameter. It is a fine-grained clay, entirely free from any coarse sand.

A chemical analysis gave the following results:—

Hygroscopic Moisture	0.99
Loss on Ignition, Combined Water, CO_2 Etc.	12.98
Combined Silica	44.92
Free Silica, or Sand	1.55
Alumina	39.13
Ferric Oxide	1.05
Lime	0.40
Magnesia	0.17
Potash	trace
Soda	trace
Total	99.20
Clay Base	97.03
Fluxing Impurities	1.62

Briquettes of this clay shrink, on drying, nine-tenths of one per cent., and have a tensile strength, dried, of 19 pounds per square inch. Its specific gravity varies from 1.55 to 1.90. On burning, there is no perceptible shrinkage. It burns to a pure white, and is apt to crackle, unless very carefully treated. It is one of the most refractory clays, tested by the writer. Its fusing point is almost equal to that of Seger cone 36.

Southeast of Lewiston, just west of the Baker estate, is an occurrence of Potomac clay, which resembles, in characteristics, that used at Stevens Pottery, Baldwin county. It is white, when dry, hard, faulted and much slickensided. In places, it is discolored with iron stains. Overlying it, unconformably, is a stratum of marl-like green-sand showing cross-bedding, and having an average thickness of six feet. This stratum contains small shell fragments in some layers. Above this, is a red and gray, laminated clay, which is overlaid, in turn, by red sand varying in thickness from four to eight feet, and grading upwards into the red, sandy clay, over which is a gray, porous sandstone, but slightly consolidated.

The clay, which is of a good quality for the manufacture of sewer-pipes, fire-brick etc., has a thickness of from fifteen to twenty feet. It is very tough, and not easily disintegrated.

Under the lens, little, if any, foreign matter is recognizable, with the exception of a reddish material, which would seem to be hematite or turgite; but, under the microscope, it appears merely as iron-stained kaolin. The clay is but slightly plastic, and does not disintegrate, on the addition of water. A high magnifying power shows many grains of quartz, which range from .001 to .004 of an inch in diameter; also, many, of smaller dimensions. The kaolin scales occur, bound together in small aggregates. Briquettes of this clay shrink, on air-drying, one eighth of one per cent., and show a tensile strength of 56 pounds per square inch. It has a specific gravity, varying from 1.67 to 1.87. This clay burns to a pure white color, with a tendency to crackle, when burned without admixture of other clays. It shrinks, on burning, an additional two per cent. of its length.

THE GORDON ROAD

On the Gordon Road, between Lewiston and Gordon, which is the next railroad station east, there is a very interesting section, exposed in a gully by the road-side. This section is as follows, in descending series: —

1. Fine, red and orange sand, with occasional partings of shale .... 80 feet
2. Clay shales, with orange sand partings .... 8 "
3. Fine white and orange sands .... 3 "

4	Alternating layers of ferruginous sand and shale	1 foot
5	Massive, gritless clay, jointed and slickensided, with many casts of shells, the shells themselves having been dissolved and removed by percolating waters. Joint planes, stained with iron and manganese	16 feet
6	Green sand and marl	4 "
7	White clay, very hard, jointed and slickensided, very pure, (grading downwards into 8)	4 "
8	Friable mixture of clay and sand, containing much sub-angular quartz	8 "
	Total	124 "

Samples of this clay were taken from four different positions in this section. One of these is a drab-colored, compact, tough clay, which breaks with a conchoidal fracture. The only impurities, revealed by the lens, are traces of muscovite. It absorbs water slightly, and does not disintegrate, when wet. Under the microscope, quartz grains appear to be almost entirely absent, and the impurities are scales and prisms of mica (the largest being .05 of an inch in diameter), and a few fragments of a colorless, slightly transparent mineral, showing a rectangular cleavage, which seems to be orthoclase. Briquettes made of this clay, shrink, on drying, *27 per cent.* in length, and have a tensile strength of 135 pounds per square inch. Its specific gravity varies from 1.83 to 2.08. It burns to a deep red color; and, although it shrinks exceedingly on drying, the additional shrinkage, in burning, is imperceptible. Its fusing point is low, being equal in refractoriness to Seger cone 26. The potash, released from the decomposing fragments of feldspar (orthoclase), are probably responsible for its comparatively low grade, in this respect.

Another sample from this locality is a brownish drab, in color,—

hard, dense and brittle, and breaking with a conchoidal fracture. It is but slightly plastic, and does not readily absorb moisture. Scales of muscovite may be seen with the naked eye; and an abundance of them, along with a few grains of transparent quartz, are visible under the lens. Under the microscope, quartz fragments are seen, some as large as .05 of an inch in diameter; also, some muscovite. Briquettes of this clay shrink, on air-drying, 25 per cent. in length. Its specific gravity varies from 1.87 to 2.00. It burns to a buff color, and it must be burned very carefully to prevent crackling. Its fusing point, as determined with the LeChatelier pyrometer, is 1,300° C.

The third sample of this clay, selected, is dense, tough and drab-colored, and has a gritty feel. On the joints, it is stained with oxides of iron and, occasionally, manganese. It contains many scales of muscovite, and quartz grains, varying, in diameter, up to one thirty-second of an inch. It absorbs water slowly. Under the microscope, quartz and muscovite are the only impurities seen. Briquettes of this clay shrink, on drying, 25 per cent. in length; and it has a tensile strength, dried, of 291 pounds, per square inch. On burning, it behaves like the last described sample.

The fourth sample resembles the third, in physical appearance. It has a specific gravity, ranging from 1.75 to 1.81, and a tensile strength, dried, of *300 pounds per square inch*. It shrinks 25 per cent., on drying, and 4 per cent., on burning.

THE WHITEHURST PROPERTY

About a mile and a half east of Lewiston, and one mile south of the railroad, on the property of J. I. Whitehurst, is an exposure, over a wide area, of the Potomac clays, which outcrop on the northern side of the hill. In the gullies, where it may be seen, this clay has a thickness of at least ten feet, and is covered with six or eight feet of stripping, which, further back in the hills, assumes an enormous thickness. The bottom of this clay is not visible. The exposure here emphasizes the fact, that the Potomac clays of this belt are not persistent in character, for any distance, but grade off rapidly into sand and sandy clays.

The quality of the clay is excellent in places; but it is frequently coarse, sandy, and somewhat discolored, all of which lessens its value. It is not unlikely, that beds of a high grade will be found on this property.

The strippings, where exposed, consist of sand and a loosely consolidated conglomerate. A sample from this locality, for laboratory examination, shows a friable, massive clay, white and pale-buff in color, and containing an abundance of muscovite scales, which may be seen with the naked eye. Under the pocket lens, occasional crystals of magnetite may be detected. Under the microscope, the amount of muscovite appears to be unusually large. Mica prisms may be seen, which are bent, and give a wavy extinction. There are, also, occasional prisms, with a high single refraction and basal cleavage, which resemble apatite. These prisms are, in all probability, prochlorite, so that there is here a mixture of two members of the mica group. Quartz is very scarce; but occasional grains of feldspar and magnetite are recognized.

This clay has a specific gravity, varying from 1.81 to 1.87. Briquettes have a linear shrinkage, on drying, of .9 of one per cent., and no further shrinkage, on burning.

Its tensile strength, dried, is 19 pounds per squre inch. Its fusing point is equal to Seger cone 35, making it one of the most refractory of clays.

THE MASSEY PROPERTY

About two and a half miles south of Gordon, which is the next railroad station east of Lewiston, is the estate of Dr. E. I. Massey. The homestead is situated on a high, narrow ridge, which is some seventeen miles in length, and has practically the same elevation, as the hills and ridges about Griswoldville.

At the base of this ridge, on the north side, are innumerable exposures of the Potomac clay. On the road, which passes the Massey house, and runs down into the valley northward, there is an interesting geological section. The descent from the house to the valley is 150 feet. For fully 120 feet, vertical distance, of this descent, the exposures are wholly red-and-orange marl-like sands, in places interbedded with thin seams of clay, and carrying fragments of the underlying clay, or "chalk." Bedding can scarcely be made out; and it seems to lie on the slope like a mantle; so that it is doubtless Lafayette. Near the top of the section, however, it is bedded, and contains clay seams; then it pitches downward, almost with the slope of the hill. On the extreme crest, the sand is vermilion-colored. At the well, near the Massey cotton-gin, and at other points of observation, these sands have a thickness of from 15 to 20 feet. The well penetrates a blue fossiliferous clay, which seems to have a thickness of 100 feet, and is probably Tertiary. It overlies the white Potomac clays and sands, at the base of the hill. The well-section does not agree with that of the road-side; but this is on account, probably, of the mantle-like nature of the sands at the latter place.

On the south side of this ridge, the overlying sands appear to have been removed; and a blue clay, which weathers white and is

POTTERY AND BRICK-WORKS OF J. W. McMILLAN, AT MILLEDGEVILLE, GEORGIA.

broken up and much slickensided, is exposed along the entire slope. It is at least 100 feet, in thickness. This bed of Tertiary materials rests unconformably on the "chalk," or Potomac clay, at the base. As mentioned before, the outcrops of the "chalk" are to be seen in gullies, in the lowlands, for miles around. At a distance, where the soil has been washed away, the effect is like that of snow-fields. The roads, too, are often colored white, by this material; and they make remarkable contrasts along-side of the red, vermilion and yellows of the sands and other clays, which are usually associated with it. The white clay, in these localities, is often entirely free from grit and sand particles. It is sometimes plastic and sometimes non-plastic, in the latter case, weathering into angular fragments. That, exposed on the Massey property, weathers and crumbles into dust. In other places, it becomes exceedingly hard on exposure, and rings under the blows of a hammer. This latter variety is characterized, when powdered, by a *mealy*, instead of the ordinary *unctuous*, feel. It contains, however, what seem to be nodules of a slightly different clay, as has been seen and noted at other localities. These so-called nodules are dark in color, and plastic. They are found in pear-like shapes, with slender necks, and look something like Prince Rupert drops, the largest part being nearly always the lowest. In places, this clay contains much iron, which is particularly common on the jointed surfaces, but which is sometimes contained in the clay, as aggregations of ochre.

The occurrence of red ochre is specially abundant on the property of Mr. N. A. Whitehurst. On the north side of this ridge, the bluish clay, which overlies the white Potomac clay, is probably the same as that, exposed on the road-side, west of Mr. Baker's house, as described above.

Samples of this clay, for laboratory purposes, were taken from four different localities. One of these may be described, macroscopically, as being a massive, fine-grained, white to buff-colored clay, which glistens brightly in the sun-shine, the light being reflected from numerous mica scales. No quartz is to be seen, under the hand lens. It absorbs water rapidly; and, when wet and moulded with the hand, no grit is felt. Under the microscope, a multitude of scales and prisms of muscovite appear, 99 per cent. of which are exceedingly minute; but a few scales range, in diameter, to .003 of an inch. Quartz seems to be absent, and magnetite crystals are very rare. It varies in specific gravity, from 1.75 to 1.85. Briquettes shrink, on drying, less than one per cent. in length; and they have a tensile strength of 23 pounds per square inch. On burning, no further shrinkage is observable. It burns white, or to a fine pinkish color, crackling more or less intensely, according to the nature and temperature of the flame, to which it is exposed. When burned carefully, it makes a smooth, porous, substantial brick, with only a faint tendency to crackle. Its fusing point is between Seger cones 35 and 36.

Another sample, taken a half-mile west of the locality, from which this clay was collected, has practically the same characteristics and properties, as do two other samples, selected at points half-a-mile apart.

The fourth sample, spoken of above, resembles these in the hand specimen; but it contains, scattered throughout, round areas of another clay, slightly rusty in color, some of which are at least two-fifths of an inch in diameter, although they are usually smaller. This clay does not rapidly absorb water; and it requires considerable rubbing, before an emulsion is obtained for microscopic study. A high power shows considerable foreign matter, varying in size up to .005 of an inch. This mineral is remarkable and unusual. It occurs in little prismatic crystals, which are sometimes straight, but often curved into semi-circles, or more complicated forms.

Such crystals have a black border, and show a rather high single refraction; but the double refraction is very weak. They range in size from mere specks, which cannot be measured, up to .0006 of an inch in diameter, .00033+ being, perhaps, the average size. In color, they are usually gray, with a faint yellowish tint. Similar bodies have been described by Ternier.[1] They have been noted in Missouri clays by Haworth, who identifies them as prochlorite. Those, which he describes, have a cleavage at right angles to the prism; while these have a cleavage parallel with the prism. The small rounded areas, spoken of, seem to differ from the rest of the clay, in that these worm-like bodies are less abundant and smaller. The clay, as a whole, is remarkably free from quartz grains; but it contains occasional needle-like crystals of rutile.

A sample from a hundred-foot bed of clay, of which the mass of the hill is here constituted, appears in the hand specimen to be fine-grained, massive, friable, and locally spotted with stains of iron oxide. No trace of foreign matter is revealed with the pocket lens. It absorbs water rapidly; becomes extremely plastic; and shows no traces of grit. Under the microscope, with a No. 3 objective, the only foreign matter appearing is an occasional scale of muscovite, or a grain of magnetite. No other impurities are revealed by a high power. It varies in specific gravity from 1.82 to 2.00. It shrinks, on drying, 9 per cent., and, on burning, an additional 2 per cent. It burns white, without crackling.

---

[1] See Compte Rendus, T. CVIII, 1889, p. 1,071.

THE R. S. SMITH PROPERTY

On the east side of the road, and just opposite the Massey property, on the estate of Mr. R. S. Smith, there are good exposures of the Potomac clay, as is also the case, on the adjoining property of Messrs. Smith and Sons.

A sample, from the latter place, shows a white, fine-grained, friable clay, specimens from near the surface being more or less stained with iron oxide. Under the lens, practically no quartz or mica is visible. A microscopic examination proves it to be a remarkably pure clay, the only foreign matter being exceedingly small prisms of muscovite, and a few specks of magnetite. The largest of the muscovite scales are .001 of an inch in diameter. Briquettes of this clay shrink, on drying, 8 per cent. of their length, and, on burning, an additional 2 per cent. Its tensile strength is 51 pounds per square inch. Its specific gravity varies from 1.55 to 1.66. It burns, without crackling, to a cream color; and its fusing point is very high, standing between that of Seger cones 35 and 36.

LOCALITIES SOUTH OF THE MASSEY PROPERTY

Traveling south and southeast from the Massey property, all the way to Big Sandy creek, which flows in an easterly direction south of Irwinton, the white Potomac clays are frequently seen in the valleys and low lands. The overlying Tertiary strata become thicker, however, as the Fall Line becomes more remote, until finally the streams fail to cut through them, and these clays remain hidden from sight, though they doubtless continue easterly and southerly far out beneath the waters of the sea and the gulf.

Samples of clay were taken from the banks of the Big Sandy, and from the properties of Messrs. Vincent and Bridges and Mr. Z. T. Miller.

THE Z. T. MILLER CLAY

The Miller property is situated about a mile, or a mile and a half, south of the Massey property. The road approaches it, by descending a steep slope, where a bed of fine clay, 120 feet thick, similar to those described at Massey's, is exposed, overlaid by orange and vermilion sands. Unconformably underlying it, is the Potomac clay. This is exposed on the Miller estate, in gullies, and at the surface in low-lying fields. The thickest exposure, yet seen, is at this locality, where at least thirty feet of the white clay is visible. It is remarkable, here, for the abundance of the pear-shaped cavities and areas, which have been spoken of above. These would seem to have originated, by the boring of worms or insects, were it not, that they permeate the whole mass, throughout its thickness, and yet do not seem to be connected, one with another. This clay is also noteworthy, on account of a property, it possesses, of hardening, on exposure to the atmosphere. Often, for several feet beneath the surface, it is exceedingly hard and tough, and is penetrated by a pick with difficulty. This property leads to a considerable use of the material, in the construction of chimneys. The soft clay is trimmed into blocks with an ax, and exposed to the weather, whereupon, they become hard and strong. Nothing, revealed by the chemical analysis, or the microscopic examination, explains this phenomenon, which may result, however, from the deposition, on drying, of an exceedingly thin film of soluble silica, among the clay particles.

A hand specimen, from the average sample collected, is cream-colored, very fine-grained and friable. The cavities, spoken of, have occasionally a dark lining of a carbonaceous material; and this suggests, as a possible explanation of their occurrence, that they originated by the decomposition, and removal by solution, of organic remains, either animal or vegetable. With the hand lens, neither quartz nor mica are discernible; and, under the microscope, no quartz can be distinguished; while muscovite is of extreme minuteness, and rare.

A chemical analysis of this clay gave the following results:—

Hygroscopic Moisture	0.21
Combined Water, Carbon Di-oxide Etc. . .	14.52
Combined Silica	42.79
Free Silica, or Sand	.82
Alumina	40.42
Ferric Oxide	.70
Potash	trace
Lime	00.37
Soda	.83
Total	100.45
Clay Base	97.73
Fluxing Impurities	1.90

This clay absorbs 80 per cent. of its weight of water. Its specific gravity ranges from 1.89 to 1.94. Briquettes shrink, on drying, 8 per cent. of their length, and an additional 4 per cent., on burning. They have a tensile strength, dried, of less than 10 pounds per square inch. It burns snowy white, with a strong tendency to crackle. Its fusing point is very high, nearly equal to that of Seger cone 36.

THE VINCENT AND BRIDGES PROPERTY

This property is one mile south of Irwinton. Clay occurs here, in much the same manner as at Miller's. It is massive, fine-grained; and, under the lens, it shows only a trace of muscovite. The microscope shows it to be one of the purest of clays. Quartz grains are very scarce. Mica, however, appears, though in small quantity; and magnetite cannot be detected. Briquettes of this clay shrink, on drying, about 9 per cent. of their length, and, on burning, an additional 4 per cent. Its tensile strength is 28 pounds per square inch, and its specific gravity varies from 1.64 to 1.73. It burns to pure white; but it cannot be utilized alone, on account of crackling. Its fusing point shows it to be one of the most refractory of clays, nearly equal to that of Seger cone 36.

THE BIG SANDY DISTRICT

Irwinton, the county-seat of Wilkinson county, is separated from Jeffersonville, the county-seat of Twiggs county, by a valley, fifteen miles broad, which is drained by Big Sandy creek. The two towns stand on ridges, of practically the same elevation; and the low land between them is much broken and dissected.

The two clay localities, last described, are typical of the many "chalk" deposits, found in this low-lying district. On Big Sandy creek, the clay has a thickness of twelve feet, the lower six feet being almost white and free from grit. The upper part of the bed is sandy and drab-colored. Overlying it, is a bed of mottled purple-and-red clay. The white variety is very hard and brittle, resembling flint, though not so hard. It can be scraped with a knife into a soapy-feeling powder. A hand specimen from an average sample of the white clay is fine-grained, compact, and breaks with a conchoidal fracture. It absorbs water quickly; but little clots seem to remain unaffected. Under the microscope, a low power shows, rarely, a small quartz grain and an occasional muscovite scale. A high power reveals only the invariable accessories — scales and prisms of muscovite.

Briquettes of this clay shrink, on drying, 6 per cent. in length, and an additional 3½ per cent., on burning. Its tensile strength, dried, is 21 pounds per square inch. Its specific gravity averages 1.66. It crackles slightly, on burning, and retains its white color, which becomes creamy, in an intense heat. It has the same high refractory quality, as that at Vincent and Bridges, and other localities.

THE McINTYRE LOCALITIES

The southwest limit of the outcrops of the Potomac clays passes Irwinton, in a northeasterly direction. McIntyre is situated on the Central Railway, about half way between Gordon and the Oconee river. In its vicinity, there are again exposed, in the low-lying lands and at the bases of the hills, the red-and-white mottled clays, which appeared so abundantly, in the general region under discussion. The following section, in descending series, is exposed, in the nu-

SEWER PIPE WORKS, MACON, GEORGIA.

merous gullies, with vertical walls from 50 to 60 feet high, on the Irwinton-Gordon road, three and a half miles from McIntyre:—

1. Red and Vermilion sands, massive, and without bedding-planes _____ 25 to 30 feet
2. Bright gold and yellow, extremely plastic clay, with thin seams of sand _____ 3 "
3. Orange and white fine sands _____ 25 to 30 "
4. White Potomac clays _____ 3 feet

The laboratory examination of the yellow, plastic clay has not been completed. It is doubtless of Tertiary age.

Two and a half miles northwest of McIntyre, at the base of the eastern slope of a ridge, there is an exposure of about forty feet of white and somewhat sandy clay, in many respects closely resembling that on the Miller property, the upper part being hard, and containing numerous areas of dark-colored, plastic clay, shrunken away from the walls of cavities. Above this, is a coarse fire-clay, overlying which are many feet of a characteristically faulted and slickensided fire-clay (the same as is described, in the notes on the Massey property), of reddish and purplish color. Overlying this fire-clay, are 60 to 75 feet of sands. The entire thickness of the section, exposed here, is 170 feet.

Just west of McIntyre, on the estate of Mr. L. A. Snow, are some outcrops of the Potomac clay, similar to other occurrences, except that, where exposed, they are somewhat discolored and sandy.

East of McIntyre, is a series of ridges and terraces, extending to the valley of the Oconee. The valley here is five miles wide, and largely occupied by low, rolling hills, similar to those in the Big Sandy valley. At frequent intervals, the chalk-like Potomac clay may be seen.

On the John Davis farm, one and a half miles east of McIntyre, it occurs with a thickness of thirty feet, the upper part being sandy. It is overlaid by a mantle, spreading down the slope of the hill, or ridge, of red and yellow Lafayette gravels, which contain numerous balls and boulders of the underlying clay. At the base of the section, the clay is almost wholly free from grit and sand; but it is largely stained with brilliant colors, red, yellow and purple. This clay would not be suitable for the demands of the paper industry; but it could be manufactured readily into paving-brick, pottery or sewer-pipe, there being plenty of fine sand, and, also, very plastic clay in the vicinity, making possible any required mixture.

Beyond this exposure, is a broad, flat-topped ridge, maintaining the general elevation, as seen at Irwinton, Gordon and Lewiston, and in the neighborhood of Griswoldville.

Laboratory experiments with this clay show, that it absorbs water and disintegrates, readily. It shrinks, on drying, 5 per cent., linear measurement, and an additional 2 per cent., on burning. Its specific gravity varies from 1.73 to 1.96; and it has a tensile strength, dried, of 21 pounds per square inch. It is apt to crackle, on burning, and it assumes a buff to fine reddish color. It fuses at about the same temperature as Seger cone 28.

On the property of Mr. August Pennington, on the flank of this ridge, there is, also, an excellent exposure of the same clay, which again out-crops in the low land at "Chalk Hill," one of the foothills on the west side of the Oconee river. The clay, at the latter place, where exposed at the surface, does not indicate a high grade. It has a velvety appearance, is gray-colored, and contains some grit. In one gully, it is seen to grade downward into the sand. It contains, also, beds of red and yellow ochre, and is stained these colors, in many places. It contains, also, the pseudo-nodules, which have been so frequently referred to.

CHALK HILL

An average sample, collected for laboratory study, shows that briquettes shrink, on drying, 4 per cent. in length, and an additional 2 per cent., on burning. Its specific gravity ranges from 1.74 to 1.90; and it has a tensile strength of 21 pounds per square inch. It burns, with little crackling, to a fine pinkish color, at temperatures obtained in ordinary kilns. It is friable and easily broken, a fact, in general, of all these Potomac clays.

On the Central Railway, a half-mile west of McIntyre, a deep red, very plastic clay is exposed in a gully. Three feet of this bed can be seen; but it is doubtless much thicker. An average sample, examined under the hand lens, reveals the pearly-lustered cleavage faces of mica plates; and, under the microscope, magnetite grains become visible, the largest of which are .0006 of an inch in diameter. Quartz is almost wholly absent. Briquettes of this clay shrink, on drying, 5 per cent. of their length, and an additional 4 per cent., on burning. Its specific gravity ranges from 1.74 to 1.85. It is almost impossible to dry it, without crackling; and its tensile strength, dried, is, therefore, not satisfactorily determined. It burns to a deep-red color, and crackles. Its fusing point has not yet been determined. It is probably low, owing to the large amount of iron oxide contained.

LOCALITIES NEAR THE MACON AND DUBLIN RAILROAD, TWIGGS COUNTY

The northern half of Twiggs county was carefully explored for Potomac and other clays, characteristic of the Fall Line belt. In addition to trips from Macon into this county, the dirt roads were traversed, with many side trips, from Griswoldville, southwesterly to Dry Branch, southeasterly, along the line of the Macon and Dublin Railroad; and thence southwesterly to Bullard's Station on the Southern Railway, and in the valley of the Oconee. From Bullard's Station, practically all of the roads were traversed north to Dry Branch, again, which is on the boundary between Bibb and Twiggs counties.

In this area, just as in the northern part of Wilkinson county, there are abundant outcrops of Potomac clay, which disappear southward, under the overlying Tertiary strata. There are also some interesting clays, in the latter formation.

THE PAYNE AND NELSON CLAY PIT

One and a half miles south of Dry Branch, and about half-a-mile northeast of the railroad, a bed of the Potomac clay has been worked by the firm of Payne and Nelson. The works, here, are of the same nature as those at Lewiston, except that operations are, at present, conducted on a smaller scale.

The clay is taken from an open pit, and is classified and stored in drying-sheds for shipment. As at Lewiston, it is generally shipped in casks, although sacks are employed for this purpose to some extent.

The stripping consists of from one to two feet of soil and gravel, and from two to six feet of the Potomac clay, which is stained with iron oxide. The total thickness of the clay is fifteen feet, the lower three feet varying from gray to drab in color, and being rather coarse and containing an abundance of magnetite crystals. The best of the clay is pure white, and resembles flour in appearance. The one defect is the occasional presence of ferruginous stains, along the joint planes, which necessitates careful sorting, and the trimming of the blocks of clay. This clay disappears, in the neighborhood, beneath the Tertiary strata, which consist of fine-grained, laminated clays, already described, as occurring at other points.

A chemical analysis gave the following results:—

Hygroscopic Moisture	1.91
Combined Water, Carbon Di-oxide Etc	13.39
Combined Silica	43.08
Free Silica, or Sand	1.94
Alumina	40.63
Ferric Oxide	1.01
Lime	0.16
Magnesia	0.00
Potash	0.27
Soda	Trace
Total	100.48
Clay Base	97.00
Fluxing Impurities	1.44

A hand specimen, from the average sample collected here, resembles the Lewiston clay. It is very friable, and varies from white to a pale-cream color. It contains an abundance of muscovite scales and some grains of magnetite, readily detected by aid of the hand lens. Under the microscope, a high power shows it to be a very pure clay, containing rare grains of quartz and magnetite, along with the muscovite, which characterizes all these clays. Briquettes of this clay shrink, on drying, 8 per cent. of their length; but, on burning, no further decrease in size is perceptible. It has a tensile strength, dried, of from 12 to 15 pounds per square inch. Its specific gravity varies from 1.72 to 1.89. It burns white, and, sometimes, to a faint pinkish color, and shows the tendency to crackle, observable in all the clays of this class. Its fusing point lies between Seger cones 35 and 36.

THE DRY BRANCH, FITZPATRICK, NAPIER'S MILL, BOND'S STORE AND BULLARD'S DEPOSITS

Going south from The Payne and Nelson Clay Works, and following the railroad, a variety of Tertiary clays may be seen in the cuts and gullies. Striking occurrences are exposed in the railroad-cut at Dry Branch, and along the highway at several places near Fitzpatrick.

A half-mile northwest of Fitzpatrick, on the Hornsby road, there is an immense gully, where the following sections are seen, in descending series:—

1	Yellow soil, grading downwards into red sand	10 to 12 feet
2	Laminated clay	5 "
3	Fine sand, with carbonaceous layers (2 to 3 inches thick)	6 "
4	Grayish, laminated clay	4 "

5 White to cream-colored clay, which dries to a hard, punky condition, is exceedingly tough, and can be cut and carved, to a remarkable extent, even when perfectly dry. It breaks with a conchoidal fracture, and the joint planes are concentrically arranged. These are stained with iron oxide, and the bed is permeated with veins of greenish-colored sands _____ 12 feet
6 Green laminated clay, very plastic and containing black carbonaceous layers _____ 8 "

About a mile west of Fitzpatrick, in a gully, by the railroad side, the clays of the above section may also be seen, to good advantage. They contain here, as at many other points in this vicinity, a few small fossil casts. The white variety was sampled for analysis and tests.

The result of the chemical analysis is as follows:—

Hygroscopic Moisture	8.70
Loss on Ignition—Combined Water, CO_2 Etc.	11.24
Soluble and Combined Silica	54.39
Free Silica, or Sand	6.89
Alumina	14.64
Ferric Oxide	0.28
Lime	7.08
Magnesia	1.71
Potash and Soda	4.23
Total	100.46
Clay Base	80.27
Fluxing Impurities	13.30

A hand specimen shows a white to grayish clay, with sometimes a yellow to greenish tint, and occasional stains of iron oxide along the joints. When dry, although exceedingly hard and tough, it cuts easily; and, when cut, it shows a smooth, glossy surface, which takes a good polish. In spite of its toughness, it is porous and almost as light as cork, floating readily on water, although, when placed in the water, it slowly moistens, becomes pasty, and, after a few minutes, sinks. Some scales of muscovite may be seen with the naked eye. The clay is not very plastic.

Under the microscope, there appears a large amount of foreign material, perhaps from fifteen to twenty per cent., which consists largely of rounded grains of quartz, the largest being .01 of an inch in diameter. Occasional magnetite grains occur, and also plates and prisms of muscovite and angular fragments of feldspar. There are also a few small areas of a yellow-colored mineral, which seems to be titanite, but which was not identified with certainty. In addition to these impurities, there occur occasional fragments of an isotropic, colored mineral, most likely basal sections of biotite. The clay particles are more or less bound together, in little aggregates. Its specific gravity varies from 0.90 to 1.20, averaging near the former amount. Experiments, with the dry, powdered clay, prove it to have an enormous absorptive-power, the clay taking up over 200 per cent. of its own weight of water. It shrinks, on drying, 25 per cent., linear measurement. A dried briquette shows an average tensile strength of 213 pounds per square inch. On burning, the dried brick shrinks an additional 6 per cent., and burns from a buff to a pale-yellow color, easily crackling and fusing, at a temperature of $1,330°$ C.

At a comparatively short distance east of the above locality, and in the vicinity of Napier's Mill, the white Potomac clay is again encountered, underlying the Tertiary beds.

A hand specimen from this locality shows a beautiful white, friable and fine-grained clay, containing an abundance of muscovite scales, with numerous quartz grains, which may be distinguished

THE CLAYS OF GEORGIA   PLATE XIV.

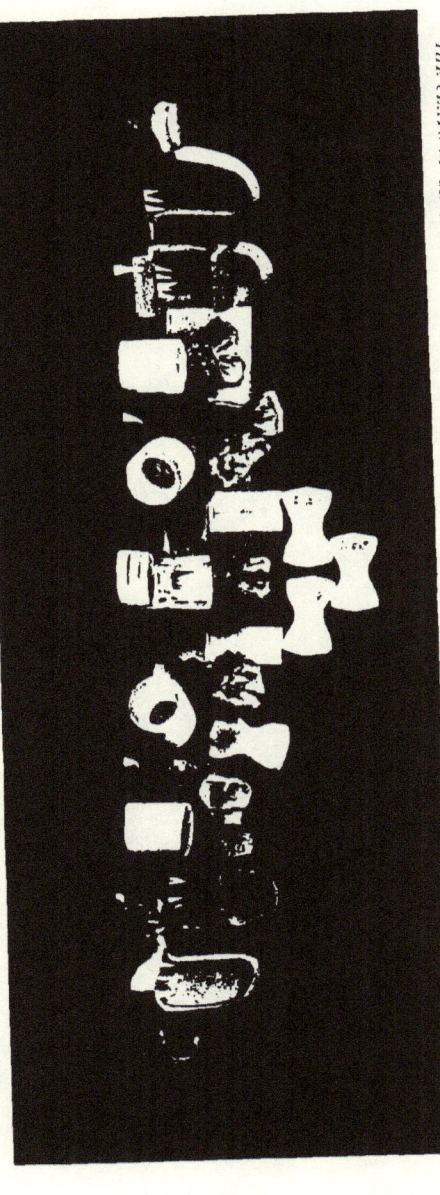

VIEW ILLUSTRATING SOME METHODS OF CLAY TESTING.

In the background, are briquettes, for tests of tensile strength, and shrinkage on drying and burning. On the sides, are types of vessels, used in testing clays and checking temperatures, in furnaces and kilns. In front, are the vessels and covers, used in testing the highest grade refractory clays. Behind these, are some, which have been through the test, and are now broken open, exposing the cones, some of which have been partially or wholly fused. In the immediate foreground, is a series of Seger cones, after a test. The unfused cone, on the left, is one inch high.

with the pocket lens. Under the microscope, muscovite, magnetite and quartz are all visible, with a low power. A high power shows further, only, that these impurities occur in great abundance. This clay shrinks, on drying, about 8 per cent., linear measurement, and but slightly more, on burning. Its tensile strength, dried, is 22 pounds per square inch. Its specific gravity varies from 1.70 to 1.85. It burns to a faint pinkish color, crackling slightly, and has a fusion point nearly equal to that of Seger cone 36.

West of Fitzpatrick, and between there and Bullard's station, there is a similar occurrence of the Potomac clay in gullies near the foot of the hills. It is here best seen at the road-side, near Bond's store, where it is in all respects similar to that at Napier's Mill.

Going east from Bond's store to Bullard's, one travels over the Tertiary uplands, and finally down steep slopes to the Columbia beds of the upper river bottom (Ocmulgee river).

THE STEVENS' POTTERY LOCALITY

Stevens' Pottery is situated in Baldwin county, on a branch of the Central Railway, running from Gordon to Milledgeville. It is so named, on account of the clay-manufacturing plant of Messrs. Stevens Brothers and Co., situated there. This firm is the largest producer of fire-clay and pottery in the State, and one of the largest producers of sewer-pipe. For a general view, see Frontispiece. This plant consists, in part, of one pipe-machine, one semidry brick-machine, three turning-wheels, two grinding-and-tempering-machines, clay-washing apparatus and compressor, two tall updraft kilns (for sewer-pipe, flues and some grades of pottery) and four circular and one rectangular down-draft kilns.

The clay, used, consists of a mixture, according to the product desired. In part, Ocmulgee alluvium from Macon is employed; and in part, red clay from the Lafayette beds on neighboring hilltops; and, further, the white fire-clay, probably Potomac, which is obtained at a number of different points on the extensive lands of this company; but chiefly, that found along the line of the railroad, for a mile or more from the works.

About three quarters of a mile west of the railroad, on the slope of a low hill, a large pit has been opened, where fire-clay is found, near the surface, and covered with but little stripping. The clay is white, as a whole; but it contains more or less iron, and is stained in places a deep red. The Lafayette gravels overlie the clay, resting unconformably on its very irregular surface. The average thickness of the latter is fourteen feet. It is somewhat sandy and micaceous. Underlying it, there is a stratum of white sand, three feet in thickness. Beneath this, in turn, is another bed of fire-clay, the thickness of which could not be determined.

An average sample of the best grade of fire-clay from this locality is very fine-grained and pale buff in color. Under the hand lens, a few scales of muscovite may be seen; but they are not abundant. Under the microscope, with the highest power, only an occasional grain of quartz can be detected; and there are no quartz grains, over .005 of an inch in diameter.

A chemical analysis gave the following results:—

Hygroscopic Moisture	0.72
Loss on Ignition—Combined Water, CO_2 Etc.	13.64
Combined Silica	43.85
Free Silica, or Sand	2.77
Aluminum	38.28
Ferric Oxide	1.02
Lime	0.18
Magnesia	0.00
Potash	0.05
Soda	0.08
Total	99.87
Clay Base	95.77
Fluxing Impurities	1.33

This clay absorbs 100 per cent. of its weight of water. It has a linear shrinkage of 8 per cent., and an additional shrinkage, on burning, of 2 per cent. Its specific gravity varies from 1.69 to 1.75; while its tensile strength is 24 pounds per square inch. It burns to a fine pinkish color, and gives a moderately firm brick. It does not crackle, if burned carefully. Its fusing point is about that of Seger cone 35.

The Lafayette, at this point, contains, in places, a clay, some of which is used, in the manufacture of brick and sewer-pipe. It has a rusty color; is porous and sandy; and contains grains of quartz, ranging up to one-sixth of an inch in diameter; also, fragments of feldspar, which have been altered to clay, since their deposition. Fully 50 per cent. of the mass consists of quartz. It seems to have been an original sandy layer, made up of quartz and feldspar, and the feldspar has in time decomposed, and furnished the clay portions, *in situ*. It is all permeated with yellow iron oxide, and it contains some large grains of magnetite.

South of the works and adjacent to the railroad track, there are several pits, or quarries, where a variety of clays may be seen. These occur at intervals, the farthest being a mile, or more, away. At many of these localities, the clay occurs at approximately the same level. At intermediate points, however, it is cut out and replaced by orange sands of the Lafayette formation.

At the first of the clay pits, going south along the railroad track, there is evidence of two unconformities. The section exposed, here, in descending series is as follows:—

1. Orange sand, containing clay in some places, and in others, coarse, water-worn pebbles. It is often indurated, and much stained with iron .... 3 to 8 feet
Unconformity.
2. White sand, which contains much kaolin resulting, in part, at least, from the decomposition of fragments of feldspar. It is indurated in places, and is often cut out by the overlying orange sands, as is also the underlying clay, which, in turn, is partly unconformably replaced .... 0 to 6 feet
Unconformity.
3. White and gray clay, which contains some sand, and is much jointed, but not so much so, as that, seen in other localities in this vicinity. It contains considerable iron oxide near the upper surface .... 3 to 6 feet
4. Cross-bedded sand, iron-stained, and both fine and coarse. It contains many thin seams of limonite, which often lie in the plains of the cross-bedding .... (?)

Resting on the irregular surface of the clay, number 3 of the section, are some interesting pebbles and boulders of clay,[1] some

---
[1] See Plates VII and VIII.

of which are several feet in diameter. They are partly, and sometimes wholly, surrounded (except at the contact surface with the white clay beneath) by an orange sand. These balls are irregular in shape, and as variable in size. They are apparently made up of small fragments of clay, water-worn, but not disintegrated, which have been re-cemented by clay. The color contrasts are sharp and the fragments stand out from the matrix beautifully. The origin of these clay-conglomerates, as they might be called, is as hard to explain, as is their present position. The writer has seen a clay breccia, where a hard flint-clay seems to have been crushed and resolidified by infiltrating clay, deposited by water; but, in the case at hand, there is too little angularity in the fragments, to suggest, that these masses, have been formed in this way, in their present position. They seem certainly to have been brought or rolled there, by currents, which cut out more or less of the clay-bed, already deposited, and which, later, deposited the surrounding materials. They are certainly younger than the clay stratum, and older than the gravels; for a stream, that could deposit such masses, would hardly deposit sands and clay.

In Plate VII, the unconformity is plainly seen, being pointed at, by the boys standing along the face of the exposure. In Plate VIII, a cross-section of the fragments of one of these boulders, three feet in diameter, is illustrated.

At the second pit, going south, a quarter of a mile, along the railroad track, still other interesting phenomena may be seen. The nature of the surface drift of orange sand changes, here, somewhat. It contains more clay and less sand; but it is remarkable for a considerable percentage of fragments of clayey shales, which it contains, a shale resembling that described, as occurring on the Gordon road, between Lewiston and Gordon, south of the Southern Railway track. In addition to these masses of shale, it contains very many irregular fragments of a white clay, evidently belonging to the older beds, lying beneath it, in the section, which is additional proof, if that were necessary, of the unconformity existing here. The white sand, found at the first pit, is wanting at the second; and, in its place, are beautiful orange sands and a bed of two varieties of clay, curiously mixed.

It is difficult to understand the relation of these two clays to each other. The one is hard and white, rubbing into a soft, flourlike material, between the fingers. It is very much broken up, and the surfaces are universally slickensided and stained with oxides of manganese and iron. Included in, or associated with, this white clay, in the same bed, are irregularly shaped masses, some a yard in diameter, of a different kind of clay, the general color of which, at a distance, is bluish. On close inspection, it appears mottled, as though some segregating action had taken place. It is quite hard, and absolutely free from grit. It cuts, with a knife, with extreme smoothness; and it is so tough, that thin, curling shavings, of considerable length, can be whittled off. It is waxy, in nature, but a little harder than wax, and submits to carving, perfectly. It would seem, therefore, to be perfectly homogeneous. Some of these masses are, however, coarser and more porous in nature. In general, on weathering, this variety breaks up, somewhat concentrically, so that, on the surface of the bed, the fragments are as small as an acorn, or smaller. As the result of this tendency, these pockets, or the masses of this particular clay, fall out from the face of the pit, faster than the white clays, surrounding them; so that they leave cavities, in the face of the latter, which is present in much the greater amount.

The phenomenon, here, is the exact opposite of that in the case of the clays, which have a nodular appearance, and which occur south of here, as described on the Miller estate, and elsewhere. In the latter case, these peculiar masses, which are, however, very different in nature, become indurated on weathering, and project above the surface.

It is difficult, at present, to account for these phenomena. It might be suggested, that the blue clay had been washed into cavities of the white, made by springs or streams of water; but it should be noted, that the white clay next to the blue clay, which is completely surrounded, is thoroughly jointed and slickensided by compressive forces. The blue clay, however, does not penetrate these cracks. It is evident, therefore, that the crushing and slickensiding of the white clay must have been later, with regard to this hypothesis, than the infiltration of the blue variety, the blue being harder, and withstanding the crushing, which broke up the white.

There are veins of the white clay, which cut across the blue masses; and this would seem to indicate, that the white clay is the younger of the two; but it is impossible to conceive, how waters, free from masses of ice, could deposit large boulders of the hard, compact clay, while it was depositing, grain by grain, the minute particles, which constitute the white clay. It is possible, of course, that the veins of white clay are secondary and comparatively recent, being subsequent to both the blue and white varieties. Near the bottom of the white clay, there seems to be a more intimate mixture of the two kinds. If these widely different clays are, then, phases of one original sedimentary bed, how can we account for the great difference in character, and, especially, that one is wholly free from grit? Different stages of induration, or the presence of varying amounts of impurities, or infiltrated mineral deposits, might easily be imagined, as occurring in different places in the same bed. Physical segregation, or flocculation, could be conceived, as bringing about striking differences in the character of such deposits in different parts of the bed. But it cannot be seen, how the phenomena here could be thus explained.

An additional matter of interest, in connection with these phenomena, is the presence, in the blue clay, of thin veins, which ramify through the boulder-like mass, but which do not penetrate the white clay, and are older than the veins of white clay, above referred to. These veins consist of what seems to be a new mineral, judging by a preliminary analysis. It is a transparent hydro-silicate of aluminum, having a yellowish color and a soft, waxy consistency.

Beneath this blue-and-white clay bed, comes in the white clay, or "chalk," where it lies at a considerably lower elevation, than at the other pits. It is highly micaceous, and is used for fire-brick and for pottery, when properly mixed with other clays. The section, here, is as follows: —

1 Orange sands, very clayey and full of fragments of shale and "chalk" .... 12 to 15 feet
2 Blue-and-white clay beds................ 8 "
3 White micaceous fire-clay............. 3+ "

A hand specimen of the fire-clay, from bed No. 3 of this section, shows a friable, white to pale-yellow clay, containing a large amount of muscovite and some small crystals of magnetite. Under the microscope, many basal sections of muscovite and magnetite, along with a few crystals of calcite, which show the planes of the rhombohedron, and are about .001 of an inch in diameter, are observable. There are also present many prismatic sections of muscovite crystals; while quartz grains are wholly absent, even those, with a diameter as small as .0002 of an inch, being rare.

This clay has a specific gravity, varying from 1.70 to 1.80. It

THE CLAYS OF GEORGIA

PLATE XV

THE FALL LINE AT COLUMBUS, GEORGIA, ON THE CHATTAHOOCHEE RIVER.   HEAD OF NAVIGATION.

shrinks, on drying, about 6 per cent.; but, on burning, no additional shrinkage is observable. It has a tensile strength, dried, of 24 pounds per square inch. It burns, with little crackling, to a white, and sometimes faint pinkish, color.

SUMMIT

About six miles northwest of Griswoldville, and north of the Fall Line, is an outlier of Tertiary strata, unlike any of those seen at other points in the State. They are exposed in the railroad cut, at a point known as "Summit," where they reach an elevation of 250 feet above the Macon bridge. These beds are limited in extent, and are surrounded by decomposed schists and diabase dikes, belonging to the Piedmont Plateau. On the eastern extension of this railroad cut, they may be seen to lie directly upon the decomposed schist, which is here cut by a dike of the fine-grained diabase rock, thirty feet in thickness.

This outlier consists of bluish and greenish, more or less laminated clay, which, in several horizons, is densely packed with shells and shell fragments. About thirty feet of clay is exposed in the railroad section; and it is capped by gravel beds of the omnipresent Lafayette.

An average sample of fossil-free material was collected and analyzed, with the following results: —

Analysis of an average sample of workable clay, collected at Summit, near Roberts station, Jones county.

Hygroscopic Moisture	3.64
Loss on Ignition, Combined Water, CO_2 Etc.	19.41
Combined Silica	13.62
Free Silica, or Sand	36.80
Alumina	11.56
Ferric Oxide	2.20
Lime	13.89
Magnesia	1.73
Potash	trace
Soda	1.36
Total	100.57
Clay Base	44.59
Fluxing Impurities	19.18

Hand specimens, taken from this bed, show a slate-colored, massive and tough clay, more or less filled with remains of gastropods. It contains, also, a few visible scales of muscovite; and it has a coarse, gritty feel.

Under the microscope, an abundance of small black specks, of what seems to be carbonaceous matter, possibly mixed with magnetite, is apparent. These vary in diameter, from .01 of an inch to .002 of an inch. Quartz grains, which range, in size, up to .007 of an inch in diameter, are seen to be present. Samples, taken from the top and bottom of this bed, respectively, show, under examination, almost identically the same characteristics, as appeared in the test of the above described sample.

This clay, when dry, absorbs 98 per cent. of its weight of water; and the dried, powdered clay, instead of shrinking and settling in the vessel, as it absorbs water through the perforated bottom, swells — in some cases, projecting an eighth of an inch, or more, above the top of the vessel.

It shrinks, on drying, 18 to 20 per cent., and, on burning, an additional 2 per cent., or even less. The tensile strength of an airdried briquette gave results, varying as follows:—

1   Sample from base of beds, 143 pounds per square inch.
2   Sample from middle of beds, 304 pounds per square inch.
3   Sample from top of beds, 255 pounds per square inch.

The specific gravity varies from 1.80 to 2.00.

Owing to the large amount of lime present, the fusing point of these clays is unusually low. The most refractory sample completely melted, at a temperature of 1,280° C. An average sample fused at 1,100°. When burned, at a moderate heat, the product is a substantial, pale-buff-colored brick, without cracks or flaws.

This clay would make an excellent material, to mix with the white Potomac clays or the alluvial clays of Macon, for the manufacture of vitrified paving-brick.

MILLEDGEVILLE

Milledgeville is situated just above the Fall Line, on the red hills, capped with the soils of the decomposed Piedmont crystallines; and it overlooks the Oconee river from the western side. The sedimentary rocks begin to disappear, about six miles south of Milledgeville, where the decomposed, upturned, micaceous schists, with their numerous quartz veins, begin to appear. For some distance north of here, however, the latter are still capped, in places, with layers of drift. In many places, the soil has lost its red color, especially near the surface, and becomes a gray or "mulatto," the change being due, apparently, to the solution of the iron stain by organic acids. The prevailing reds, in the gneiss and schist regions, are probably due to the presence of iron-bearing minerals, which furnish a constant supply, through their decomposition, of the iron, which is oxidized and stains the soil.

In the city of Milledgeville, there is but one industry, utilizing clays, namely, The Milledgeville Brick and Pottery Works, the property of Mr. J. W. McMillan. These are the only clay-works in Baldwin county, except those at Stevens' Pottery. They are situated on the second river-bottom of the Oconee, near the foot-bridge, which crosses that river.

The clay, which is worked here, is doubtless of Columbia age. It is a plastic alluvium, which varies in character, as all alluvial clays do; but it is workable, at this point, to a depth of ten feet from the surface. With the clay, obtained locally, is mixed, to some extent, for the manufacture of bricks, the white, micaceous Potomac clay, obtained at Carr's station, in Hancock county, where it occurs in a cut on the Georgia Railroad. The clays are mixed, according to the desired texture and color of the burned product.

The product consists of common building-brick, re-pressed face- and ornamental-brick, and, also, tiles and a variety of pottery, such as garden urns, fancy vases, and large jugs and flower-pots, the latter being of exceptional high-grade. About 2,000 bricks per day are made. The plant, as illustrated on Plate XII, consists of an office, store-house, drying-house, Penfield brick-machine, hand re-press-machine, turning wheels, flower-pot machine, with a capacity of 3,500 per day, five rectangular brick-kilns, one circular, up-draft pottery-kiln, divided into vertical compartments, one rectangular, down-draft pottery-kiln, one steam-engine and boilers, and a tram and tram-cars.

The works were established in 1887; and the product is shipped all over the South, even to points in Texas.

Wood is used in burning; but, at the time of the writer's visit, it was the intention of the proprietor to use coal, and put in new and improved machinery.

THE REGION WEST OF MILLEDGEVILLE

West of Milledgeville, along the Fall Line towards Macon, the character of the crystalline rocks varies rapidly; and they pass from coarse gneisses, resembling granite, but with a true gneissic, banded structure, to laminated schists, the planes of schistosity being folded, dipping downward, here and there, and giving the rocks the appearance of standing on edge. They are decomposed, to a depth of from 15 to 20 feet, but not completely so; as there are, usually, in a decomposed mass, cores of solid rock. The color of the soil varies, as already stated, from red to gray; and there is often a sudden transition in color, corresponding to the kind of crystalline rock below. Among the sedimentary formations south of the Fall Line, there are also red and gray soils, sharply separated from each other, in just the same way; though the sudden transition in this case is the result of a different cause.

THE AUGUSTA REGION AND EASTERN EXTENSION OF THE FALL LINE CLAYS OF GEORGIA

East of Baldwin county, the Fall Line, as stated on preceding pages, crosses and re-crosses the Georgia Railroad, on the whole averaging south of it. The white Potomac clay occurs, as described in the Milledgeville notes, at, or near, Carr's station on the Georgia Railroad, in Hancock county, and at other points, on this road, near Augusta. South of here, its occurrence becomes rare, until it finally disappears, as usual, beneath the Tertiary strata. It may be found, at intervals, along the lines of the Augusta Southern Railroad, between Sandersville and Augusta, and crossing parts of Washington, Glascock, Jefferson and Richmond counties.

THE WORTHEN LOCALITY

The most important exposures, along the line of the Augusta Southern Railway, are near Chalker and Worthen, in Washington county. It occurs, in large quantities, at Magnolia Plains on Floyd's creek, and at the Worthen and Upper Mills on Keg creek. At the first of these three places, it is pure white, and contains but little grit. At the other, it is coarser and more siliceous. Washington county, in which these occurrences are situated, has a number of small potteries, operated in a manner, similar to those in Crawford county. The wares are made, at intervals during the year, and are disposed of, by peddling through the country.

The leading potteries are those of J. M. Bussell, John Redfern and Andrew Redfern. The business has been carried on, for the past 75 years, or more.

At Chalker, the white clay, which seems to be Potomac, is exposed in the railroad cut, and has a very irregular surface, showing a maximum thickness of twelve feet, and being unconformably overlaid by the following series of Tertiary and Lafayette material: —

1. Lafayette, sandy loam, covering the slope } ---- 15 to 20 feet
2. Coarse, micaceous, laminated clay, varying in color from drab to yellow, weathered into small cubical and angular fragments .... 6 "
3. Fossiliferous, sandy clay, indurated and siliceous in places .... 4 to 6 "
4. Hard, siliceous layers, undermined by the stream in the immediate vicinity, where beds are exposed thirty feet in thickness and resembling a quartzite. It contains many fossils .... 1 to 2½ "
5. Dark-drab and brown, iron-stained, sandy clay .... 8 "

THE AUGUSTA CLAYS

The geological conditions, seen at Augusta, which is situated at the head of navigation on the Savannah river, are very much like those at Macon and Columbus, more especially Macon, where the Potomac formation is not observable in the immediate vicinity of the city.

West of the city, however, and at a considerable elevation above it, in Richmond and Columbia counties, are exposures of what is doubtless the white Potomac clay. In this region, there are, also, observable numerous sections of Tertiary strata; abundant sections of the Lafayette, resting unconformably on the strata beneath it; and frequent sections of the decomposed crystallines, showing the unconformable contact between them and the overlying sediments, which contain, at their base, numerous fragments of the crystalline rocks, and immense quantities of muscovite and other minerals derived from it.

THE CLAY INDUSTRY OF AUGUSTA

The clay industry of Augusta is confined chiefly to the manufacture of building-brick, although a small amount of tiles, ornamental ware and terra-cotta is produced. A part of the industry, really belonging to this community, is across the river, in South Carolina.

The clay, used, is Columbia alluvium, occurring in the plains, on which the business portion of Augusta stands. The maximum production on both sides of the river, in any one year, has been 30,000,000 bricks; while the average annual production is about 20,000,000.

THE AUGUSTA BRICK COMPANY

This company is said to have the largest brick-yard in Georgia. It was established by a man named DeLaigo, a refugee from the San Domingo massacre. The product consists of common and

THE CLAYS OF GEORGIA

PLATE XVI

GENERAL VIEW OF THE FALL LINE, OPPOSITE COLUMBUS, GEORGIA, SHOWING RECENT EROSION. THE UPTURNED CRYSTALLINE ROCKS ARE OVERLAID BY CRETACEOUS AND COLUMBIA FORMATIONS. (For Detail View, See Plate XVII.)

ornamental brick (hand pressed). The average annual production is said to be about 14,000,000, most of which is common brick. The bricks are made by 3 Sword machines (stiff clay process), having a capacity of 10,000 per day. They are burned, in four ordinary clamp kilns. Wood, at the average cost of $3.00 per cord, is used in preference to coal, for burning. The plant is situated on the southeastern edge of Augusta, on the flood-plain, and about 30 feet above the surface of the river. The clay used is excellent in quality, and free from any excess of sand and mica. A sample of alluvial clay, a fair average of that used by the various companies, was collected for physical examination, the results of which were as follows: —

Its specific gravity is approximately 2.00. It shrinks, on drying, 8 per cent., linear measurement. Its tensile strength, dried, is 148 pounds per square inch. It burns to a fine, strong, red brick; but, owing to the impurities present, it is, of course, not a refractory or fire brick. It fuses, at a little less than 1,300° C.

THE WILBUR BOSWELL PLANT

Mr. Boswell produces, annually, with a Sword machine and four or five clamp kilns, a large number of common- and a few face-brick. His plant has a capacity of 30,000 per day.

THE McCOY BRICK AND TILE COMPANY

This company has an elaborately equipped and thoroughly modern brick-yard. The plant consists, in part, of one continuous kiln, of the Youngrens pattern, with a capacity of 30,000 per day of twenty-four hours; one Swift kiln, changed from an original Morrison, with a capacity of 500,000; one Phillips dryer (with ten racks), heated by exhaust steam from the engine, which is run through 160,000 feet of pipe, the air, thus heated, being agitated by 26 fans. The bricks are dried, in the last named apparatus, in about 24 hours. They are introduced and removed, on cars. It has a capacity of 20,000 per day of ten hours. There are also open-dryers, consisting of covered racks, which carry 120,000 bricks, and are capable of turning out 20,000 per day.

The system of operating, at these works, is interesting. The clay is brought from the dump, in cars holding a little less than a cubic foot each. These are elevated, on cables, by a winding-drum, and are emptied into an immense granulator, which crushes and mixes the clay with sand. This granulator delivers the clay, at an angle of 45°, into a machine, which is crusher, pug-mill and auger brick-machine combined, and which delivers, upon an apron, side-cut brick, with the "flats," rough, and the exposed edges, smooth. The machine is made by E. M. Freese & Co., of Galion, Ohio. It has a guaranteed capacity of 40,000 per day, and can be made to produce 6,000 per hour. The company has, also, a 100 horse-power engine, with 110 horse-power boilers, and a 55,000-gallon water-tank, elevated on a 72-foot tower. The yard is equipped for re-press work; but, with the brick-machine, now used, this process is not necessary.

The clay is obtained from the river bottom, about 3/8 of a mile from the works, to which it is drawn, on a tram, by mules.

THE GROVETOWN DISTRICT

Grovetown is about 20 miles west of Augusta, on the Georgia Railroad, near the southeastern boundary of Columbia county. Here, as also in Berzelia on the west, Belair on the east, and the adjacent territory in Richmond county on the south, there occurs a great abundance of clays of many varieties.

The white clay, or "chalk," has been shipped from a number of points in this region, for a variety of uses. The only developments of clay, in these two counties, except that of the industries, already described in Augusta, is that of a small pottery at Grovetown, where, convenient to the railroad and an excellent bed of clay, a high grade of common pottery is produced.

In a railroad cut, just east of Grovetown station, there is an exposure, for a considerable distance, of the upper part of a bed of clay, only three feet of which can be seen, which is remarkable in character. When fresh from the bed, it is from gray to buff in color; but it is sometimes stained with yellow and reddish tints. It seems to be wholly free from grit, and is carved readily with a knife; but it will not disintegrate in water, although it can be crumbled with the fingers; and, when ground up, it becomes exceedingly plastic. It is a soft clay; but it breaks, on weathering, concentrically. Although soft, neither rains, nor the waters, which constantly seep through its joint-planes, and flow down into the railroad ditch, make any appreciable impression on the bed, as regards the removal of material.

In spite of its seemingly favorable properties, it cannot be used, at least without the addition of other clays, in the manufacture of burned products, owing to its tendency to pop and fly to pieces, with explosive violence, on being heated, regardless of the care, with which this is done. The cause of this has not yet been investigated. Fossils are said to have been found in this bed; but none were seen by the writer. It may be of Cretaceous age; but it is probably Tertiary.

A hand specimen of this clay may be described as being from buff to gray in color, with yellow iron-oxide stains, along the joints. It is compact and very fine-grained, and has a conchoidal fracture. Under the pocket lens, it shows traces of mica, but no quartz. Under the microscope, with a high power, minute scales of muscovite appear in great abundance; also quartz grains, ranging up to .001 of an inch in diameter, and an occasional minute crystal of magnetite. The specific gravity of this clay is found to be very uniform throughout the bed. The determination of many samples gave 1.30 as a result, with no appreciable variation. It shrinks, on drying, about 16 per cent., and, on burning, an additional 4 per cent. It is one of the lightest clays examined. The tensile strength of dried briquettes is 144 pounds per square inch. It burns to a buff color; but, however carefully burned, it bursts into many fragments, at a comparatively low temperature.

It is overlaid, with a bed of Lafayette red clayey gravel, which would make an excellent road material, and might be profitably used for that purpose in Augusta.

SILAS REED'S ESTATE

On the property of Mr. Silas Reed, near Grovetown, a small stream, resulting from springs, has caused a "break," a quarter of a mile in length, along the hillside. The water moves on the surface of a bed of dark-colored, tough clay, known in this locality, as "umber," and once worked for a paint material. It is simply a clay, which contains a large amount of carbonaceous material, in places becoming so far lignitic in character, that it will burn, although with difficulty. Sometimes the clay resembles that, described as occurring in the railroad cut at Grovetown; and, also, the exceedingly light variety found at Fitzpatrick. Dr. Hatton, of Grovetown, stated to the writer, that fossils have been found in this horizon, where it has been penetrated by wells.

This bed of lignite and clay is exposed to a depth of from 15 to 18 feet. Its color varies from cream to brown and black. A deep shaft was sunk here, by prospectors in search of umber; but a record of the beds passed through could not be obtained. This clay does not absorb water, at least until crushed and powdered. Under the microscope, small amounts of muscovite, occasional grains of quartz, and large quantities of organic matter are visible. Overlying it, is an enormous mass of red, deep-orange, and golden-yellow sands, with, here and there, a seam of clay, evidently of Tertiary age.

THE MARY INGLETT PROPERTY

This property is about three miles south of the Georgia Railroad, near Grovetown. There is here, in a gully, on the edge of a sandy plateau, an exposure of white clay, known locally by the usual name, "chalk." It is almost wholly buried, and hard to examine. Its thickness could not be ascertained. The clay and conditions in general suggest strongly the occurrence on the VanBuren estate in Jones county, as described above. The clay bed is overlaid by a sand, mostly quite coarse, and consisting of fragments of white and smoky quartz, and containing partings and occasional isolated masses of clay, derived from the bed beneath. The surface of the country, about, is covered with just such sand, the smaller grains of which are sharp and angular, while the larger are well rounded. This formation is many colored; contains, throughout, seams of clay; and is doubtless Lafayette. On the extreme surface, many of the quartz pebbles have been carved by wind-blown sand to a remarkable extent. Among these, have been found good imitations of octahedrons; and those, showing pyramids on one side, are of frequent occurrence. The clay itself is not all white; but it is occasionally pink and mottled. The pocket lens reveals an abundance of quartz grains and many scales of mica. Some of the quartz is smoky. There may also be detected many distinct cleavage fragments of partially kaolinized feldspar. Under the microscope, the quartz grains appear in great abundance, and many scales and prisms of muscovite are visible. There may, also, be seen a few remarkably shaped bodies, the dimensions of which vary from .01 to .02 of an inch. They are probably organic, and most likely, the remains of diatoms. Prisms of zircon, also, occur. The clay has a specific gravity, varying from 1.8 to 1.9. It shrinks, on drying, 8 per cent., linear measurement, and an additional 2 per cent., on burning. It has a tensile strength, dried, of 20 pounds per square inch. It burns to a pinkish or cream color. Its fusing point corresponds to that of Seger cone 30; and it is therefore highly refractory.

BELAIR

There is an excellent opportunity at this place, to study the relations between the decomposed crystalline rocks and the overlying sediments, which are so largely derived from them. It is well illustrated by a section in the railroad cut, just east of the station. The exposure, here, is about 20 feet in height. The fissile structure of the decomposed schists, at the base of the section, is very pronounced; although the decomposition of the rock, as a whole, is almost complete; and the resulting material is, at least in many places, homogeneous, fine-grained and plastic. It is beautifully vari-colored, — chocolate, red, golden-yellow, white and gray being the predominating colors.

In some localities, this rock seems to have been a fine-grained chloritic schist; but, in other places, it resembles shale. The dip of the schistose planes varies from 45° to 60°.

This crystalline rock is cut by prominent quartz veins, which, although broken, have remained hard and unaltered, while the rock about it is decomposed. The quartz veins project above the irregular surface of the crystallines, showing, that, during the process of erosion, which went on, when this was a land surface, the surrounding rock was worn away, further than the quartz veins.

Another feature, which tells the same geological story, is the distribution, at the contact of the sediments with the upturned crystallines, or along the whole irregular surface of the latter, of large irregular-shaped boulders of quartz, which are evidently residuary from quartz veins, once resting in the crystalline rocks, but which were left behind, as surface boulders, while the bed-rock itself was decomposed and removed. About these, but of course unconformably resting on the crystallines, is a bed of coarse angular sand and gravel.

Since the clays, which have been studied, have been largely derived, with their impurities, from the decomposed crystallines, a sample of the latter, from this place (where, although it has not been removed and sorted out by the running water, it is nevertheless thoroughly decomposed), was selected and submitted to physical tests, similar to those employed in the case of clays.

The results of a physical examination of this material were as follows:— A hand specimen shows a massive mixture of kaolinite, muscovite, quartz and feldspar, and also numerous bright crystals of magnetite. It is, therefore, seen to contain exactly the same minerals, that are found in all the clays examined along the Fall Line, except that the latter consists more completely of the finer-grained material, kaolin, the coarse particles, quartz, magnetite, feldspar and muscovite having been removed by the differential carrying capacity of moving water. Under the microscope, with a low power, a large amount of sericite, in minute crystal plates, and occasional crystals of calcite are seen. The specific gravity of this material varies from 2.03 to 2.09. It shrinks, on drying, 4 per cent., linear measurement; but, on burning, there is no additional shrinkage. The tensile strength of the dried mass is very low, being not over five or six pounds per square inch; and the dried mass crumbles readily between the fingers. It burns to a porous, friable substance, and would make good brick material, with the addition of a little plastic clay. It is not refractory, but fuses completely, at a comparatively low temperature.

H. F. CAMPFIELD'S PROPERTY

A section, somewhat similar to this one at Belair, may be seen on the property of Mr. H. F. Campfield, about three quarters of a mile east of Lula station, on the Knoxville road.

About ten yards northwest of the exposure in the railroad cut at Belair, there is an occurrence of sand and gravel, sometimes white, but often red and yellow, and containing occasional thin seams of

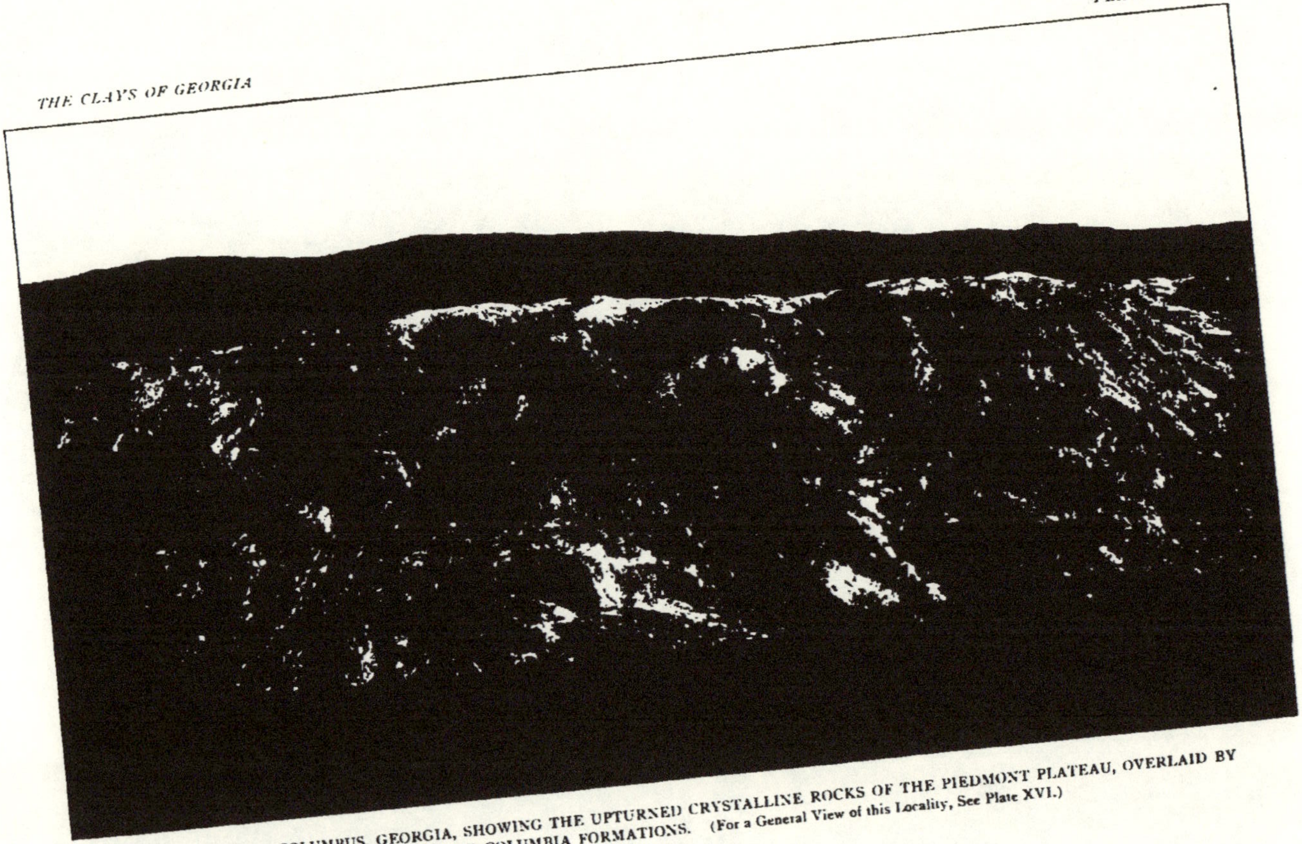

THE CLAYS OF GEORGIA

PLATE XVII

VIEW OPPOSITE COLUMBUS, GEORGIA, SHOWING THE UPTURNED CRYSTALLINE ROCKS OF THE PIEDMONT PLATEAU, OVERLAID BY CRETACEOUS AND COLUMBIA FORMATIONS. (For a General View of this Locality, See Plate XVI.)

limonite; while, here and there, it is cemented into a solid conglomerate by iron oxide. Contained in these gravels are, also, irregular-shaped pockets, sometimes several feet in thickness, of fire-clay, which dries white, although it is frequently colored with iron-oxide stains. It is fine-grained.

Beds of fire-clay, similar to those occurring in the gravels at this locality, are found at other points in this region, notably at a point on the railroad, near the 12-mile post (between Belair and Grovetown), and, also, at a point 2½ miles south of Belair, where there is a bed of it, excellent in quality.

Partially buried beneath this gravel, which may be either Tertiary or Lafayette, is a comparatively hard rock, which seems to be a partially decomposed gneiss. The structural planes are, however, horizontal; and it may consist of slightly re-arranged arkose, derived from gneisses, which contain an abundance of quartz and feldspar crystals, the latter, in this case, being thoroughly kaolinized and consequently soft. Such beds as this need be observed critically, to distinguish them from many of the sedimentary gravels, which overlie them.

A hand specimen of the decomposed gneiss, shows a white to gray, porous-looking material, stratified or laminated, and very friable. It is gritty and possesses little plasticity. In a strong light, it glitters, with a white, satin-like luster, caused by an immense number of minute scales of some micaceous mineral. Under the microscope, grains of quartz are visible, ranging up to .01 of an inch in diameter. Fully 50 per cent. of the mass is seen to consist of prisms and plates of a pale-straw-colored mica, the plates ranging up to .003 + of an inch, in diameter. These plates, or scales, appear to be muscovite, and the prismatic sections give very high interference colors. It might be classified as sericite. The material contains also a little magnetite and masses of kaolinite scales.

DEPOSITS EAST OF BELAIR, ALONG THE GEORGIA RAILROAD

East of Belair, at the eight-mile post, on the Georgia Railroad, are thick and massive beds of clay, sand and gravel, the clay, or kaolin, being often intimately mixed with the gravel, in a manner hard to explain, as it would seem, that water, depositing such coarse sand and gravel, which is often cross-bedded, thus showing the presence of strong and varying currents, would entirely wash away the fine-grained clay.

On both sides of the cut, striking unconformities are observable. An illustration of the better of these is shown in Plate V. The irregular surface of the lower-lying series seems to have been carved beneath flowing waters, which, subsequently, under changed conditions, deposited the overburden. The contours of the surface of the former are too sharp and irregular, to have been developed on a land surface. There is probably represented, here, a deposit of Lafayette on the Lower Cretaceous mixture of sand and gravels, the former consisting of materials derived from the latter, which is apparently a continuation of the sands occurring in Jones, Baldwin and other counties, under similar conditions.

In the high slope, above this exposure, are coarse beds of residuary gravel, lying on the extreme surface. They are mostly composed of smooth, well-rounded pebbles. The fields, too, are often covered, after a heavy rain, with very coarse, white and transparent quartz sands, which produce a glittering effect like that of snow.

NINETEEN-MILE POST, GEORGIA RAILROAD
RICHMOND COUNTY

At this locality, there is a clay, almost white in color, which is very plastic, and contains isolated grains of quartz, scattered through the mass. It has been crushed, and again cemented together, by the infiltration of a finer clay. The upper layer of the bed, two feet in thickness, is full of areas of clay, with the peculiar nodular effect, characteristic of the variety found on the Miller estate, which has been described above. These areas are somewhat indurated, as in the case of the Miller clay; and, where the bed has been weathered, they project beyond the surface. Unconformably overlying this clay, is a bed of grayish and yellow, very hard, but rather porous, clay, containing sand, in patches and isolated particles, in all, 15 to 18 feet in thickness. It is vari-colored, being gray, yellow and pink in different places, and frequently mottled. The quartz grains, contained in it, are rounded. It contains the same filled cavities, or nodules (?), as in the bed below. Overlying it, are massive beds of incoherent bands, which, however, in places, are strongly cross-bedded and full of water-worn pebbles, some of which are rounded fragments of limonite. These sands carry, also, much clay, scattered through them, and as numerous vari-colored partings.

A hand specimen from *the lower bed* shows grains of quartz, some of which are as much as a sixteenth of an inch in diameter. Under the microscope, grains of magnetite are visible, and some quartz, although small grains of this mineral are rare, the larger ones being such as could be easily removed by washing. Muscovite is in abundance, there being only an occasional prism, but what is over .001 of an inch in length. The kaolin is in aggregates, which do not readily disintegrate. Zircon needles are observable, averaging .003 of an inch in diameter. The little aggregates of kaolin vary in size from .001 of an inch to .0002 in diameter; and they are usually stained a pale-brown color.

A hand specimen, selected from *the upper bed*, is of a beautiful pink color, with numerous parallel partings, varying in color, and showing a thin seam of iron oxide deposited along the planes of separation. Considerable mica can be recognized with a pocket lens. The study of this clay, under the microscope, reveals a considerable amount of foreign matter, consisting mostly of angular quartz grains, the largest of which are .0006 of an inch in diameter. With a high power, the double-refracting clay particles are numerous; but nearly all are so small, that their thickness is insufficient to produce an interference color of the first order. The finest material is often stained a pale-yellow color. Muscovite is present, in basal sections, in prisms up to .0008 of an inch in diameter.

The specific gravity of this clay varies from 1.53 to 1.76. It shrinks, on drying, about 10 per cent., linear measurement, and an additional 5 per cent., on burning. Its tensile strength, dried, is 25 pounds per square inch. It assumes a pink to reddish color, on burning, and makes a porous, friable brick, when burned without admixture of other clay.

THE SEVENTEEN-MILE POST, GEORGIA RAILROAD
COLUMBIA COUNTY

At the 17-mile post, there occurs a white to yellow friable clay, with comparatively little grit, and free from impurities of any kind.

It resembles the best clay, seen in Jones and adjacent counties. Its specific gravity is 1.70. It shrinks, on drying, 8 per cent., and a small additional per cent., on burning. Its tensile strength, dried, is 25 pounds per square inch. It is highly refractory, and has a fusing point between that of Seger cones 35 and 36, and is, therefore, one of the best fire-clays, as regards refractoriness, which has been examined.

ALONG THE GEORGIA RAILROAD, WEST OF GROVETOWN

Traveling westward on the Georgia Railroad from Grovetown, clays are observed in cuts, much like those already described, in this vicinity, being colored red, white and yellow. They are seen, also, near Boneville, just east of the station, and, again, some little distance west of it.

The layer of sediments near Boneville, becomes very thin, and the underlying schists are exposed, in all the stream-beds. Near Thomson, granitic rocks occur; and, from Thomson to Camack, the crystalline rocks alone are exposed.

CHAPTER IV

A COMPARISON BETWEEN A GEORGIA CLAY AND OTHER WELL-KNOWN CLAYS OF THE UNITED STATES

The localities of a series of widely-known clays, which the writer has selected, and which are here placed in comparison with one of the Georgia clays examined, were personally visited by him, in connection with a study of the Clays of the United States, begun several years ago; and the samples, from each locality, were personally and carefully collected, with the view to obtaining, as nearly as possible, an average, for the given locality, of the workable part of each clay-bed, under consideration. The experiments and analyses were, also, personally conducted.

THE DEFINITION, OCCURRENCE AND DISTRIBUTION OF FIRE-CLAY

The term, *fire-clay*, is generally used, from its commercial significance, as a name for clays, which have refractory qualities, sufficient to enable them to resist the high temperatures, used in the technical arts, mostly in the construction of furnaces and of vessels, in which glass is to be made, zinc smelted, ores assayed, etc.

A pure clay is highly infusible; and the clays, selected for comparison with the Georgia clay, are those, which contain such small quantities of fluxing impurities, as to enable them to withstand high temperatures. As the clay occurs in nature, the lack of these impurities, detrimental to it as a fire-clay, may be due, either to their original absence, or to the fact, that they have been dissolved and removed, by natural processes. All clays probably originate, in the presence of these impurities; and transported clays may have, deposited with them, more or less of such; and it is probably due, in most cases, at least, to a final removal of deleterious compounds, that clays of high refractoriness exist. The conditions, favorable for the removal of the impurities, are largely, that the clay-bed shall be exposed, for a long time, to the influence of comparatively pure waters, charged with some solvent reagents, and shall be sufficiently porous, at least in some period of its existence, to allow the free percolation and escape of these waters.

GEOLOGICAL AND GEOGRAPHICAL OCCURRENCE

Clays, which have been so situated, occur in many parts of the world, and have been commercially developed, largely in Great Britain, Germany, France and the United States of America, — less extensively in Sweden and other countries.

In the United States, they occur in localities, scattered over the length and breadth of the country. Geologically, they are practically confined to the Coal Measures, where they are most largely developed, and to the Cretaceous and Tertiary formations. The most extensively worked of these is the Allegany Coal Measure basin; and, within its boundaries, the greatest development of the fire-clay industries is in Pennsylvania, Ohio and Maryland. The Illinois Coal basin is worked, in a number of localities, over a wide area. The outlying portion of this basin in St. Louis county, Missouri, is enormously developed within the city-limits of St. Louis, mostly in the vicinity of the suburb, Cheltenham. The coal basin, which extends southwestward from Iowa, furnishes an abundance of good fire-clay, and is most largely worked within the limits of Missouri.

The Cretaceous formations furnish high-grade refractory clays, which are worked very extensively in the State of New Jersey, and in the neighboring portions of New York, on Staten and Long Islands. Beds of the Cretaceous are also worked very successfully in the States of Colorado and Georgia.

The Tertiary formations are worked for fire-clays in the States of Texas, California and Florida.

In one district, namely, that developed on the Kentucky and Ohio sides of the Ohio river, near Portsmouth, Ohio, a bed of fire-clay is worked, which belongs to the Lower Carboniferous period.

Fire-clays in the United States are worked, and form the basis of important industries, in New York, New Jersey, Pennsylvania, Ohio, Maryland, Georgia, Kentucky, Tennessee, Illinois, Indiana, Iowa, Missouri, Texas, Colorado and California; and it is probable, that, if fire-clays are not already mined in many other States, they soon will be. Among these are Michigan, West Virginia, North and South Carolina, Alabama and Arkansas.

CLAYS SELECTED FOR SPECIAL STUDY

The clays, discussed in this chapter, were selected, on account of their significance, as furnishing, to some extent, typical samples from the most important fire-clay beds, when considered from the point of view of their commercial value, geological age, and geographical position. They come from the following localities:—

Woodbridge, New Jersey	St. Louis, Missouri
Woodland, Pennsylvania	Athens, Texas
Mount Savage, Maryland	Golden, Colorado
Canal Dover, Ohio	Carbondale, California
Sciotoville, "	Griswoldville, Georgia

NEW JERSEY

This State has long been noted for its clay industries. The center of development of these is in Middlesex county,[1] which lies just west of Staten Island Sound and Raritan Bay.

The clay-beds in the Middlesex district belong to the Cretaceous formation, and occur at the bottom of the New Jersey Cretaceous series, where they are known as the Plastic Clays, a synonym of "Potomac," "Tuscaloosa" and "Trinity," terms used along the southwestward continuation of this formation. The Cretaceous strata occur, in surface distribution, as a broad belt, stretching

---

[1] The geology of this region, the nature and origin of the clays, and the industries based on them, have been most admirably studied and described by the late Professor George H. Cook, State Geologist, and his assistant, Mr. John C. Smock, since State Geologist, in a report, already referred to, which was published by the Geological Survey of New Jersey in 1878.

across the State from Sandy Hook to Delaware Bay, having a northwesterly strike, and dipping gently to the southeast, beneath the Tertiary strata, and out under the Atlantic ocean. The Plastic Clay beds are lowest down in the series; and, consequently, they outcrop farthest west. There are but comparatively few places, where these clays actually outcrop at the surface, since they are often buried under many feet of sand and gravel drift.

South of the region under discussion, within New Jersey, the clay-beds are so completely covered by drift, as to be little exploited, or worked.

Clays, belonging to this same formation, extend northeastward, and are worked, for commercial purposes, on Staten Island and Long Island, New York, and on Martha's Vineyard at Gay Head, in Massachusetts. In the two former places, they are mined for the manufacture of fire-brick; but, in the latter, for pottery and tile-making. The Southern extension of these clays furnishes material for many clay industries in a number of States, particularly in South Carolina and Georgia.

In the Plastic Clays, there are three beds, which, although they are used commercially, for various purposes, are designated as "fire-clay," and furnish excellent material for the manufacture of refractory wares. These are the South Amboy, the Woodbridge and the Raritan fire-clay beds. A sample of clay from the bed at Woodbridge, was taken by the writer, as before stated; and the results from his study of it may be seen in the several comparative tables, near the end of this chapter.

MARYLAND[1]

Maryland is the seat of numerous and important clay industries. Brick, terra-cotta, tile and pottery are manufactured largely, in the vicinity of Baltimore, where the clay is furnished by the Potomac and Columbia formations. Fire-clay is also mined, from these formations, at Baltimore, and at North East, in Cecil county. The most important fire-clay locality is, however, in the western part of the State, where the clay is obtained from the Coal Measures. The development of the clay is confined to Allegany county; and the locality, in general, is known as the Mount Savage Fire-clay district. The general geology of this district is closely related to that of Western Pennsylvania and Eastern Ohio.

The Mount Savage Fire-clay district is situated in the western part of Maryland, and southwestern part of Pennsylvania. It is one part of the Carboniferous field, which extends, as already outlined, from Northeastern Pennsylvania, through Eastern Ohio, Maryland, West Virginia, Kentucky and Tennessee, into Alabama. Within this broader area, a large number of fire-clay beds are worked. Few of them, however, are continuous over any great extent of country; and none of them approach uniformity, in character and quality, throughout any considerable distance. Some beds are more persistent than others, in all these respects; and, among them, a few furnish exceptionally good clay. These have been developed mostly in Western Pennsylvania, Eastern Ohio, the northern part

---

[1] The geological literature and notes of these fire-clays are confined essentially to various reports of the Pennsylvania and Ohio Geological Surveys; to a paper by the late Professor George H. Williams and others, in a book entitled "Maryland, Its Resources, Industries and Institutions," published for the Chicago Exposition; and to a paper on the manufacture of fire-brick at Mount Savage, Maryland, by Robert Anderson Cook, in Vol. XIV, Trans. Am. Inst. Min. Eng.

of Central Kentucky, and in Allegany, Maryland. They have been slightly developed in West Virginia, Tennessee and Alabama.

The "Mount Savage bed," proper, is worked in Allegany county, Maryland; in Bedford and Somerset counties, Pennsylvania, immediately north; and, it is said,[1] in Cambria county, Pennsylvania, at Johnstown. The numerous other fire-clays worked, north and west of this region, appear to be in other beds in the geological scale.

The fire-clays belong to the group, originally designated, in the Paleozoic formations, as No. XII, by Rogers, and known as the Pottsville or Seral Conglomerate. It belongs, geologically, in this region, at the bottom of the Coal Measures, just beneath the Lower Productive Measures, and above the Mauch Chunk Shales.

The earliest and best known of the fire-clay works, in this region, are those at Mount Savage. Other works, which mine the same bed, are at Frostburg, Maryland, and Hyndman, Bedford county, and Keystone Junction, Somerset county, Pennsylvania; possibly, also, at Williams station in the last named county. As before mentioned, the writer visited Mount Savage, and selected, from its clay bed, a sample for study, the results of which may be seen in the tables of comparison, near the end of this chapter.

PENNSYLVANIA

Pennsylvania is one of the leading fire-clay-producing States in the Union. Kaolin, *in situ*, is mined in the southeastern part of the State, and the *transported* clays, of several geological ages, in a

---

[1] See report H H, 2nd Geol. Surv. of Penn., 1875, p. 99.

great many localities. The best known fire-clays are those, occurring in the Coal Measures, in the western part of the State. Beds, occurring at different horizons, are worked at different places.

The clay, underlying the Freeport Upper Coal, exists in numerous places, as an excellent fire-clay, and has been mined extensively at numerous localities, the best known of which is Bolivar. It is also worked at several places in Fayette, Indiana, Beaver and Westmoreland counties.

The Lower Kittanning Coal is often underlaid by a fire-clay, which is also extensively mined for a high-grade refractory material, at a number of places, both in Pennsylvania and Ohio. The workable part of the bed is usually six or seven feet thick, and the whole bed varies in thickness, from ten to fifteen feet. It is mined in Fayette and Westmoreland, and, probably, in other counties. In some counties, it is said, that the Ferriferous Limestone Clay is worked as a fire-clay. The plastic clay, occurring immediately below the Clarion Coal-bed, is mined at Bolivar, on the Conemaugh river, and is used in admixture with non-plastic clays, in the manufacture of refractory materials. The Brookville coal under-clay (?) is mined at several places in Clearfield county; and, among other localities, at Brookville, in Jefferson county; Queen's Run and Farrandsville, in Clinton county; and Johnstown, in Cambria county.

The two best known fire-clay districts are the Clearfield and the Mount Savage, which, as was stated in the notes on Maryland, extend into Pennsylvania. The results of the writer's investigation of the sample, taken from the clay bed at Woodland, Clearfield county, are to be seen in the comparison tables, near the end of this chapter.

OHIO

The clays of this State have been carefully studied by Professor Edward Orton, State Geologist of Ohio, and Mr. Edward Orton, Jr., who have made elaborate and comprehensive reports on their distribution, origin, nature and uses.[1]

Clays for pottery, tile, sewer-pipe and brick-making are mined from the beds, belonging to several geological periods, at many different points in the State. The Clays of the Coal Measures, however, vastly exceed in importance those of all the other formations, and the so-called Kittanning Clays and Shales are the all-important clays of this series.

The beds, which furnish the highest grade fire-clays, and are accessible, are limited, with few exceptions, to two districts — one, in the northeastern part of the State, covered largely by Tuscarawas county; the other, in the central part of the southern margin of the State, in Scioto county, which, considered as a clay district, extends across the Ohio river into Green and Carter counties, Kentucky. In the Tuscarawas county region, the fire-clay, worked, is the Lower Kittanning Under-clay, previously mentioned in the notes on the Pennsylvania fire-clays. The occurrence of this clay has been admirably described by Mr. Orton, as follows:—

"The first district of Kittanning clays, met, on entering this State from the east, is the yellow-ware beds of Liverpool and Wellsville. The vein, here, is plastic and easily mined, and is reached mainly through drifts. It underlies the Kittanning Coal, directly.

The next district is the Jefferson County pipe-works, which line the Ohio river, for twenty miles, from Wellsville nearly to Steuben-

---

[1] Vol. V, Geol. Surv. of Ohio, Report on Economic Geology. Ibid., Vol. VII, Part I, 1893.

ville. The clay is here regularly excellent in quality, easily found, and at convenient levels for mining. It is rather sandy, or siliceous, quite hard at first, but quickly slacking on exposure; and it is plastic and readily moulded, after grinding. The clay extends up Yellow creek, beyond Irondale, where it still is of good quality, and is in constant use. In the northern part of Tuscarawas county, at Mineral Point, the Kittanning clay appears, in a new and valuable phase. It is found, as a very hard and fine fire-clay, suitable for making any refractory material. It is used for retorts, glass-pots, brick of finest special grades, etc. It is found, about three feet below the Kittanning coal, separated from it by worthless clay. There can scarcely be a more beautiful or faultless stratum of clay, than the hard clay of this place. It lies in a band, three and one half feet thick, showing faces smoother, than the most regular coal; it is so hard, that chips from a pick-blow will cut the hands of the miner. It is of a light-drab tint, burning to a light-cream color; it shows the blue, concentric venation of organic matter, noticed in the flint-clay of Sciotoville. The mining of this clay is carried along with the mining of coal, both being taken out of the same pit-mouth. The coal is usually worked first, and then the clay. There is enough plastic clay occurring with the flint, to allow its perfect working. This bed of hard clay marks the Kittanning horizon from Magnolia."

The Scioto county clay belongs beneath the Maxville, or Sub-carboniferous Limestone, coal (?), and is the lowest fire-clay bed, in the Geological scale, worked in this country. This clay has been described by Mr. Orton,[1] as follows:—

"The deposit is first found in the tops of intersecting ridges near Portsmouth, occasionally overlain by nodules of limestone, but no

---

[1] Economic Geology of Ohio, Vol. V, "Clay Deposits," page 661.

coal; at Sciotoville, eight miles eastward, the hills are all high enough to hold it, and its best development is here; at Webster, thirteen miles further on, it is at drainage-level; and, shortly after, dips under the surface. On the Kentucky shore, and for forty miles to the southeast, it has a very heavy development. The clay runs, from one foot to five feet thick, averaging two feet, six inches. It is benched, where practicable; drifted, where necessary; and all is mined by powder. It is grayish-drab in color, full of blue organic stains, but of remarkable purity and excellence. It is very hard and flinty; but it runs soft in places, so that plastic clay need not, as a rule, be imported.

"Another small development of this clay is found near Logan, Hocking county, where it supports the manufacture of a good firebrick."

Two samples of clay from this State were investigated, by the writer. They were taken, respectively, from Canal Dover and Sciotoville. The results may be seen in the tables of comparison, near the close of this chapter.

MISSOURI

Missouri completes a quintette of States, with those already discussed, remarkable for their clay deposits. These abound within her borders, in endless variety, and are suitable for nearly all the uses, to which clay is put. Her fire-clays are wide-spread and varied, in manner of occurrence. They may be grouped for convenience of study, according to their geological occurrence, into three classes, which may be designated as the St. Louis, Crawford and Audrain Clays.

The first and most important of these classes includes those, which are mined in the city of St. Louis, and which belong geologically to the Coal Measure basin, lying in St. Louis county and city, with the exception of a small area in St. Charles county, just across the Missouri river. This Coal Measure basin may be considered as an outlier of the Illinois Coal basin, since it is geologically nearer that basin, than the one lying west of it. It will be discussed more in detail, below.

The next class, the Crawford clay, includes a large number of more or less isolated pockets of fire-clay, occurring, unconformably, in cavities and along old valleys, among the Silurian and, possibly in some cases, Devonian and Lower Carboniferous rocks. These pockets of fire-clay are distributed over a number of counties, occupying about the center of the eastern half of the State, its northern boundary being but a few miles north of the Missouri river. They are of uncertain geological age, and are possibly the remnants of a once very extensive formation in this region. They are to be found mostly in the minor lateral valleys, along the borders of the greater ones, and apparently always near the tops of the valley sides. They sometimes occur in shallow pockets along the top of the divides, this being especially noticeable in Gasconade county, where they occur over a wide area. The greatest thickness of the clay, seen by the writer, was at Regina, Jefferson county, where a pocket had been opened, to a depth of sixty feet. Borings have been made in pockets, seeming to belong to this class, which penetrated the clay to a depth of 125 feet.

The clay is usually of a cream-color; but it is often mottled with purple and reddish tints, which are organic stains, and which readily disappear, on ignition. It is hard and brittle; breaks with a conchoidal fracture; and weathers concentrically, breaking up indefinitely into sharp, angular fragments. It is mined (as a fire-clay) mostly in Montgomery, Warren, Franklin, Crawford and Phelps counties, for shipment only, going to fire-brick works in St. Louis, Chicago and eastern cities, where it is used in connection with more plastic clays, to diminish their shrinkage.

In one locality, near Union in Franklin county, the upper four or five feet of this clay is plastic. In a great many other places, it occurs as white clay, brilliantly mottled with reddish tints, and sometimes stained very dark purple. It is comparatively soft, and free from sand, and has a smooth, soapy feel; and it is cut with a knife, in an indescribably smooth, soft way.

The third class, Audrain clay, most highly developed in Audrain and Callaway counties, belongs in the Coal Measure basin, extending southwestward from Iowa, across Missouri, and occupying the greater part of the northwestern half of the State. The workable fire-clays of this class are confined, largely, to the southeastern border of this area. They were developed at the time of the writer's study of this region, in 1890, only at Fulton, Mexico, Ladonia and Vandalia. The Fulton clay (in Callaway county) may possibly belong to the second class, described above. At Mexico, this fire-clay is largely developed, as it is, also, at Vandalia; but it is not certain, that the clay at these two places belongs to the same bed.

ST. LOUIS, MISSOURI

The St. Louis fire-clays are, by far, the best known in Missouri. They are the only ones from the State, that are represented in this comparative study.

The following descriptive notes of the St. Louis locality and its clays are reproduced, largely, from a report, published by the writer in 1891.[1]

Geologically speaking, St. Louis has three local sources of clay. These are the Coal Measures, the Quaternary deposits, and the residuary product of decomposing limestone. Of the two latter, the Quaternary deposits only are of much value; and these, not being used as fire-clay, will not be discussed.

The Coal Measure Clays

The Coal Measures, which have an area, approximating two hundred square miles in St. Louis county and city, carry practically an inexhaustible supply of potter's, sewer-pipe and fire clays, of which the best of the latter have long been noted for their high refractory qualities. The mining and manufacturing of these clays has been carried on, for many years, and, to a large extent, in the immediate neighborhood of St. Louis.

The fire-clays occur in beds, ranging from a few inches to seven, or more, feet in thickness. They are occasionally so near the surface, as to be mined by drifts; but usually they have to be sought by shafts. The deepest of these in St. Louis is 120 feet, from the surface of the ground to the bottom of the clay bed.

These clays, even in the same bed, vary largely, in character and composition. They are, usually, grayish in color, and very hard, when mined; but, after exposure to the weather for a variable time, the clay softens, or slacks, and falls into a loose mass of finely divided particles. Pyrite occurs in the clay, generally in aggregations, and either close to the top or bottom of the clay bed, so that it may be avoided, to a great extent, in mining. The best grades of clay, those suitable for making glass-house pots, in a raw and

---

[1] Bulletin No. 3, Geological Survey of Missouri.

unwashed condition, are of limited occurrence, and are not continuous in the bed, in which they are found. They may be considered as existing in "pockets"; but the transition, from the best clay into the poorest grades, is gradual.

The fire-clay is mined in chambers, which are kept well timbered, until they have been carried as far as possible, when the timbers are removed, and the roof is allowed to drop in, the débris forming a wall for a new chamber, which will be run along the side of the old one. There are probably several miles of these chambers within the city of St. Louis. Immediately overlying the clay, is, generally, a hard shale or a bed of sandstone, which makes a very good roof. At some of the mines a bed of coal lies just above the fire-clay; but it rapidly thins out, and almost quite disappears, at other mines. The use of powder is a necessity in the mining, on account of the hardness and density of the clay.

Two samples from the St. Louis fire-clays were taken, respectively, from the Evens & Howard mines and the factory of the Christy Fire-clay Co., the sample from the latter being of their washed pot-clay. The tables of comparison, near the end of this chapter, show the results of the writer's investigations.

COLORADO

The Colorado clays are representative, in a general way, of the clays, occurring in the Eastern Rocky Mountains. They belong mainly to the Cretaceous and Paleozoic formations, the best and most used being the Upper Cretaceous. These clays have been so little prospected and developed, in other regions in this part of the country, that no comparison can be made; and it cannot be asserted, although it is probable, that clays, belonging to the same horizon as those worked for refractory materials in Colorado, occur of equal purity elsewhere.

The fire-clays of this State furnish, largely, the refractory material for the furnaces in the Rocky Mountain region. They are taken from the Dakota member of the Upper Cretaceous formation, which flanks, with its upturned strata, the eastern margin of the mountains, and makes the "bordering reefs," locally known as "hog-backs."

The sample taken from the fire-clay bed at Golden, was studied by the writer, and his results are included in the tables of comparison, near the end of this chapter.

CALIFORNIA

Clays are mined in a number of places, in different parts of the State; but little has been published, concerning their geological occurrence and age. They, probably, are mostly from the Tertiary formations, which are distributed in the Sacramento and San Joaquin valleys and the minor valleys of the Coast Range. These clays are used largely, for the manufacture of pottery, sewer-pipe etc.; and but one locality has been found to produce a refractory clay. This is in Amador county, along the foot-hills, and near the eastern border of the Tertiary strata, which occupy the great valley of California, and which are thought to be of Pliocene age. The clay is mined, in the vicinities of Ione and Carbondale, part of it being locally manufactured into pottery, but most of it being shipped to Sacramento and San Francisco, where it is manufactured into fire-brick and various other products.

The sample, taken by the writer, from the clay banks at Carbondale, was submitted to investigation, the results of which are given in the comparison tables.

TEXAS

Texas, like California, has geological formations, furnishing an abundance of clay material, which has been but little worked. The most important of the clay-producing formations, omitting from consideration the possible value of kaolin deposits among the granitic rocks, are the Cretaceous and Tertiary, which occupy the greater part of the State. The most valuable clays, however, are confined to eastern localities, and belong to the Miocene, or Fayette, division of the Tertiary. While beds, from this geological horizon, are mined at a number of places, for pottery and the commoner uses, there are only three localities, where they are worked as fire-clay. These are in Henderson, Limestone and Fayette counties, the beds worked in the two latter counties being assigned to the so-called Timber Belt beds.[1]

The Fayette clays are referred to, in a report by the State Geologist, as follows:—[2]

"The Fayette beds contain light-colored clays, many of which are pure white. These beds of clay not only underlie and overlie the middle beds of Fayette sands; but they are also found interbedded with that series."

Many analyses have been made of clays of various portions of

---

[1] Second An. Rep., State Geologist of Texas, p. 74.

[2] Excerpt, "Clay Materials of the U. S.," by Robert T. Hill, page 521.

these beds. While some of them are too high in alkalies, or fusible constituents, others are well suited, for the manufacture of all grades of earthenware, below that of porcelain, or French china, as it is called. Clays of this variety have been secured, in various localities, from Angelina to, and below, Fayette county.

The most important of the fire-clay localities, in Texas, is that at Athens, the county-seat of Henderson county, in the northeast part of the State. At this point, the clay belongs to the Miocene, or Fayette, horizon. The results of the writer's investigations of a sample, taken from the bed at Athens, may be seen in the tables of comparison.

GEORGIA

All the geological formations of this State furnish clays, either residual or sedimentary; and, in many parts of the State, these are found of sufficient purity, to warrant their use as refractory material. Clays, for other purposes, occur abundantly, and are widely distributed. They are used in the manufacture of all kinds of building-brick, both red and buff, sewer-pipes, pottery, terra-cotta, fire-brick and fire-proofing.

The clays of the Fall Line, from which the typical fire-clay sample is selected, have already been discussed in another chapter,[1] and the facts will be but briefly summarized, here.

The fire-clay horizon belongs to the Potomac division of the Lower Cretaceous, which extends across the State, in a narrow belt, running from Columbus to Augusta, and is largely overlaid by the sands and clays of the Tertiary, Lafayette or Columbia formations.

---

[1] See Chapter II.

The locality, from which the type-clay, used in this comparative study, was selected, is Griswoldville, Jones county, which is located near the center of the most extensive and most developed outcroppings of the high-grade Potomac clays in Georgia. This clay is used, at many localities, in the manufacture of pottery, usually on a small scale; and, at only one place, namely, Stevens Pottery, in the manufacture of fire-brick. Stevens Pottery is but a few miles northeast of Griswoldville. At Lewiston, about six miles east, and at Dry Branch, about five miles west of Griswoldville, this same clay is mined for shipment to manufacturers of wall-paper. The clay is found in all directions from Griswoldville, being exposed in railroad cuts, or in gullies, in the field. It is mined in open cuts, being sometimes bare of stripping; but, usually, it is covered, by from eight to twenty feet of sand and gravel. For the analysis and detailed description of this clay, the reader is referred to Chapter III, pages 107–109 inclusive, where they occur, in the notes on the Fall Line Clays.

TABLE OF CHEMICAL ANALYSES[1]

Name of Clay	Typical Georgia Clay	Woodbridge	Canal Dover	Woodland	Christy	Carbondale	Golden	Evens & Howard	Sciotoville	Athens	Mt. Savage
Water Lost, on Drying at 100° C., or "Hygroscopic Moisture"	0.57	1.34	1.23	1.08	2.63	2.42	0.83	2.74	1.40	1.82	1.43
Combined Water, CO_2 and Organic Matter	13.08	14.03	10.22	12.89	8.94	12.76	13.30	10.20	13.11	7.17	12.62
Combined Silica, SiO_2	44.94	43.46	32.03	44.84	26.03	36.03	45.99	27.56	43.28	31.82	41.93
Alumina, Al_2O_3	39.13	36.98	28.14	37.27	21.16	31.75	31.72	23.26	40.04	20.71	38.14
Free Silica, or Sand	1.23	1.92	26.81	0.61	38.32	17.00	5.22	31.82	0.31	37.06	5.90
Ferric Oxide, Fe_2O_3	0.45	0.99	1.67	1.48	2.72	1.06	0.36	3.24	2.40	1.01	1.31
Lime, CaO	0.18	0.15	0.49	0.58	0.61	0.87	0.36	0.65	0.15	0.22	trace
Magnesia, MgO	0.11	0.83	0.28	0.62	0.30	0.04	0.23	0.42	0.17	0.39	0.17
Potash, K_2O	0.51	0.21	0.15	0.89	0.86	0.16	0.48	0.54	0.30	0.69	0.49
Soda, Na_2O	0.63	0.28	0.46	0.19	0.88	0.48	0.45	0.79	0.38	0.39	0.42
Total	100.45	98.75	100.25	99.37	99.82[3]	100.15	98.11	98.48[4]	100.14	99.46	100.98
Clay Base[2]	97.34	94.37	70.39	95.00	56.13	80.54	91.01	61.02	96.43	59.70	92.69
Fluxing Impurities	1.88	2.46	3.45	3.74	5.37	2.58	2.27	5.64	3.40	2.70	2.39

[1] Based on Clay, dried at 100° C.　　[2] Approximate.
[3] Contained, also, Sulphur, .12, and Sulphur Tri-oxide, .56.　　[4] Contained, also, Sulphur, .31, and Sulphur Tri-oxide, .35.

TABLES OF COMPARISON

Clay Base Present

1	Griswoldville, Georgia	97.34	per cent.
2	Sciotoville, Ohio	96.43	" "
3	Woodland, Penn.	95.00	" "
4	Woodbridge, N. J.	94.37	" "
5	Mount Savage, Md.	92.69	" "
6	Golden, Colo.	91.01	" "
7	Carbondale, Cal.	80.54	" "
8	Canal Dover, Ohio	70.39	" "
9	Evens & Howard, St. Louis, Mo.	61.02	" "
10	Athens, Texas	59.70	" "
11	Christy, St. Louis, Mo.	56.13	" "

Sand Present

1	Sciotoville, Ohio	0.31	per cent.
2	Woodland, Penn.	0.61	" "
3	Griswoldville, Georgia	1.23	" "
4	Woodbridge, N. J.	1.92	" "
5	Golden, Colo.	5.22	" "
6	Mount Savage, Md.	5.90	" "
7	Carbondale, Cal.	17.00	" "
8	Canal Dover, Ohio	26.81	" "
9	Evens & Howard, St. Louis, Mo.	31.82	" "
10	Athens, Texas	37.06	" "
11	Christy, St. Louis, Mo.	38.32	" "

Fluxes Present

1	Griswoldville, Georgia	1.88 per cent.
2	Golden, Colo.	2.27 " "
3	Mount Savage, Md.	2.39 " "
4	Woodbridge, N. J.	2.46 " "
5	Carbondale, Cal.	2.58 " "
6	Athens, Texas	2.70 " "
7	Canal Dover, Ohio	3.45 " "
8	Sciotoville, "	3.40 " "
9	Woodland, Penn.	3.74 " "
10	Christy, St. Louis, Mo.	5.37 " "
11	Evens & Howard, St. Louis, Mo.	5.64 " "

Scale of Relative Densities

(No. 1, Least Dense)

1. Griswoldville, Georgia
2. Woodbridge, N. J.
3. Athens, Texas
4. Carbondale, Cal.
5. Canal Dover, Ohio
6. Golden, Colo.
7. Mount Savage, Md.
8. Sciotoville, Ohio
9. Woodland, Penn.
10. Evens & Howard, St. Louis, Mo.
11. Christy, St. Louis, Mo.

A COMPARISON OF CLAYS

Relative Fusibility [1]

(No. 1, Least Fusible)

		Seger Cone
1	Griswoldville, Georgia	36/35
2	Woodbridge, N. J.	35/34
3	Woodland, Penn.	34/35
4	Mount Savage, Md.	34/35
5	Sciotoville, Ohio	34
6	Carbondale, Cal.	33–34
7	Christy, St. Louis, Mo.	33
8	Evens & Howard, St. Louis, Mo.	32/33
9	Canal Dover, Ohio	32
10	Golden, Colo.	32/31
11	Athens, Texas	29

Percentage Amount of Water Absorbed

1	Griswoldville, Georgia	85.00
2	Carbondale, Cal.	84.00
3	Woodbridge, N. J.	75.00
4	Evens & Howard, St. Louis, Mo.	73.00
5	Athens, Texas	70.00
6	Golden, Colo.	65.00
7	Woodland, Penn.	54.00
8	Sciotoville, Ohio	53.00
9	Canal Dover, Ohio	53.00
10	Christy, St. Louis, Mo.	52.00
11	Mount Savage, Md.	46.00

---

[1] The form of expression "34/35" means, that the clay corresponds more closely to cone 34, than to cone 35. The form "33-34" means, that the refractoriness of the clay is half-way between that of cone 33 and cone 34.

A COMPARISON OF CLAYS

Shrinkage on Drying

(No. 1, Least)

1. Canal Dover, Ohio .... 8.00 per cent.
2. Sciotoville, Ohio .... 8.00 " "
3. Carbondale, Cal. .... 8.05 " "
4. Woodland, Penn. .... 8.12 " "
5. Griswoldville, Georgia .... 8.30 " "
6. Golden, Colo. .... 8.30 " "
7. Mount Savage, Md. .... 8.35 " "
8. Christy, St. Louis, Mo. .... 12.30 " "
9. Athens, Texas .... 12.40 " "
10. Evens & Howard, St. Louis, Mo. 12.40 " "
11. Woodbridge, N. J. .... 12.50 " "

Additional Shrinkage on Burning

1. Griswoldville, Georgia .... 0.0 per cent.
2. Athens, Texas .... 0.0 " "
3. Sciotoville, Ohio .... 0.3 " "
4. Evens & Howard, St. Louis, Mo. 0.3 " "
5. Christy, St. Louis, Mo. .... 0.5 " "
6. Carbondale, Cal. .... 0.8 " "
7. Golden, Colo. .... 1.5 " "
8. Canal Dover, Ohio .... 2.3 " "
9. Woodland, Penn. .... 3.1 " "
10. Mount Savage, Md. .... 3.7 " "
11. Woodbridge, N. J. .... 4.4 " "

Tensile Strength

(Wet)

1. Christy, St. Louis, Mo. — *Strongest*
2. Woodbridge, N. J.
3. Athens, Texas
4. Woodland, Penn.
5. Evens & Howard, St. Louis, Mo.
6. Griswoldville, Georgia
7. Carbondale, Cal.
8. Sciotoville, Ohio
9. Canal Dover, "
10. Golden, Colo.
11. Mount Savage, Md. — *Weakest*

Tensile Strength

(Dry)

1. Christry, St. Louis, Mo. — 156 lbs. per sq. in.
2. Evens & Howard, St. Louis, Mo. 128 " " " "
3. Athens, Texas — 115 " " " "
4. Golden, Colo. — 60 " " " "
5. Mt. Savage, Md. — 49 " " " "
6. Canal Dover, Ohio — 47 " " " "
7. Sciotoville, Ohio — 38 " " " "
8. Griswoldville, Ga. — 32 " " " "
9. Woodland, Penn. — 31 " " " "
10. Woodbridge, N. J. — 24 " " " "
11. Carbondale, Cal. — 16 " " " "

Scale of Penetrability

(Wet)

No. 1, Least Penetrated

1. Mount Savage, Md.
2. Athens, Texas
3. Christy, St. Louis, Mo.
4. Woodbridge, N. J.
5. Golden, Colo.
6. Griswoldville, Georgia
7. Carbondale, Cal.
8. Evens & Howard, St. Louis, Mo.
9. Canal Dover, Ohio
10. Sciotoville, Ohio
11. Woodland, Penn.

Scale of Relative Hardness

(As Clay Occurs in the Field)

1. Carbondale, Cal. — *Softest*
2. Athens, Texas
3. Griswoldville, Georgia
4. Woodbridge, N. J.
5. Christy, St. Louis, Mo.
6. Evens & Howard, St. Louis, Mo.
7. Golden, Colo.
8. Sciotoville, Ohio
9. Canal Dover, "
10. Woodland, Penn.
11. Mount Savage, Md. — *Hardest*

The foregoing tables show, of the Georgia fire-clay, that, of the whole series, it is the most refractory, and contains the largest amount of clay base, the smallest amount of fluxing impurities and nearly the smallest amount of sand. It is one of the softest clays in the series; is the least dense; and absorbs the largest percentage of water. As regards shrinkage on drying, it stands intermediate in the series; but there is no further shrinkage on burning, which is true of only one other sample in the series, namely, that from Athens, Texas. In tensile strength, both dry and wet, and in penetrability, it is intermediate. In most of the important properties for a fire-clay, it is remarkable. There are few fire-clays, however, which have such a range of properties, as makes them suitable for manufacture, without a mixture of other clays; as, for instance, to make them more plastic, less plastic, more refractory, shrink less, or possess a greater tensile strength and "body." Its one defect is, that it crackles, on burning; but this can be easily remedied.

APPENDIX

By W. S. YEATES, State Geologist

Dr. Ladd having failed to furnish the "list of publications," referred to, on page 36, I have compiled a bibliography of such publications on the subject of Clay and its manufacture, as seemed desirable for this work. For an excellent and very extensive bibliography, the reader is referred to the work of Dr. J. C. Branner, mentioned in the list below.

BIBLIOGRAPHY

Anderson, F. Paul. Paving Bricks. Tested for Compressive Strength and Absorption. The Digest of Physical Tests and Laboratory Practice, Vol. I, No. 2, pp. 108–133. Philadelphia, April, 1896.
Barber, Edwin A. The Pottery and Porcelain of the United States. New York, 1893.
Barus, Carl. Thermal Effects of the Action of Aqueous Vapor on Feldspathic Rocks (Kaolinization). School of Mines Quarterly, Vol. VI, No. 1, pp. 1–231.
──────── Kaolinization (in Geology of the Comstock Lode and the Washoe District, Mont.) U. S. Geol. Surv., Vol. III, pp. 290-308. Washington, 1882.
──────── Subsidence of Fine Solid Particles in Liquids. Bull., U. S. Geol. Surv., No. 36. Washington, 1887.

Becker, G. F.   Geology of the Comstock Lode.   Mon., U. S. Geol. Surv., Vol. III, Washington, 1882. (Decomposition of Feldspar, p. 385; Kaolinization, p. 388; Clays, p. 394; Thermal Effect of Kaolinization, pp. 397-400.).

Bischof, C.   Die Feuerfesten Thone.   Leipzig, 1895.

Blatchley, W. S.   A Preliminary Report on the Clays and Clay Industries of the Coal-bearing Counties of Indiana. 20th An. Rep. (1895) of the Dept. of Geology, Indiana, pp. 24-179. Indianapolis, 1896.

Blue, A.   Vitrified Bricks for Pavements.   3rd An. Rep., Ontario Bureau of Mines, p. 103.   Toronto, 1893.

Branner, J. C.   Bibliography of Clays and the Ceramic Arts.   Bull., U. S. Geol. Surv., No. 143, pp. 1-114.   Washington, 1896.

Brewer, W. H.   On the Suspension and Sedimentation of Clays. Am. Jour. Sci., 3rd Ser., Vol. XXIX, pp. 1-5. 1885. Abstr., *Jarbuch für Mineral.*, Vol. I, p. 414. 1888.

Calvin, Samuel.   Geol. of Allamakee County, Iowa.   3rd An. Rep., Iowa Geol. Surv., 1894. Des Moines, 1895. (Clays, pp. 94-96.)

———   Geology of Jones County, Iowa.   Iowa Geol. Surv., An. Rep., Vol. V, 1895. Des Moines, 1896. (Clays, pp. 107-109.).

Chamberlin, T. C.   Color of Milwaukee Bricks and Analyses of Clays.   Geol. Rep. of Wisconsin, Vol. I, p. 120; Vol. II., pp. 235, 236.

——— **and Salisbury, R. D.**   The Driftless Area of the Upper Mississippi.   6th An. Rep., U. S. Geol. Surv., for the year 1884-85, pp. 199-322. Washington, 1885. (Analyses of Residual and Glacial Clays, p. 250.).

Chase, Charles P.   Paving Bricks and Brick Pavements.   Engineering News, Vol. XXIV, July 19, p. 55; July 26, p. 70, 1890.

Clarke, F. W.   (Analyses of Eight) Clays, etc., from Martha's Vineyard, Mass. Bull., U. S. Geol. Surv., No. 55, pp. 89-90. Washington, 1889.

Collett, John.   Kaolin (in Owen County, Indiana).   7th An. Rep., State Geologist, p. 358. Indianapolis, 1876.

———   Clays (of Indiana).   11th An. Rep., State Geologist, p. 21. Indianapolis, 1882.

———   Clays and Kaolin (in Indiana).   12th An. Rep., State Geologist, p. 24. Indianapolis, 1883.

Cook, Geo. H. Clays of New Jersey. Geol. Surv., New Jersey, pp. 48–69. 1878.

Cook, R. A. The Manufacture of Fire-brick at Mt. Savage, Md. Trans., Am. Inst. Min. Eng., Vol. XIV, p. 701. Eng. News, Vol. XV, p. 227, April 10, 1886. Eng. and Min. Jour., New York, March 13, 1886.

Cox, E. T. Porcelain, Tile and Potters Clays. Indiana Geol. Surv., p. 154. 1878.

Davis, Charles Thomas. A Practical Treatise on the Manufacture of Bricks, Tiles, Terra-cotta etc. Philadelphia, 1884; 2nd ed., 1889.

Day, David T. Pottery. Mineral Resources, U. S., 1889–90, pp. 441–444.

Day, Wm. C. Structural Materials. Mineral Resources, U. S., 1886. (Bricks etc., pp. 556–580.)

————— Structural Materials. Mineral Resources, U. S., 1887. (Bricks etc., pp. 534–551.)

————— Structural Materials. Mineral Resources, U. S., 1888. (Bricks etc., pp. 557–575.)

Engle, G. B., Jr. Enameled Bricks. Am. Arch., Vol. XLIII, p. 129. 1894.

Ferry, C. Fire-clays in New Jersey. The Iron Age, Vol. LV, p. 332.

————— High Silica Fire-brick. The Iron Age, Oct. 10, 1894.

————— Value of Practical Tests of Refractoriness. The Iron Age, Oct. 31, 1894.

Griffiths, H. H. Clay Glazes and Enamels. Indianapolis, 1895.

Harden, E. B. Report on Fire-clay. An. Rep., Geol. Surv., Pennsylvania, 1885, pp. 239–249.

Hay, Robert. The River Counties of Kansas. Trans., Kan. Acad. Sci., Vol. XIV, 1893–'94. Topeka, 1896. (Clays, pp. 243–245; 252–253; 258.).

Henderson, J. T. Commonwealth of Georgia, Atlanta, 1885. (Clay, p. 132.).

Hill, Robert T. Clay Materials of the United States. Mineral Resources, U. S., 1891, pp. 474–528; the same for 1892, pp. 712–738; the same for 1893, pp. 603–617. Washington, 1893, 1893, 1894.

Hofman, H. O. Further Experiments for Determining the Fusibility of Fire-clays. Trans., Am. Inst. Min. Eng., Vol. XXV, 1895; also Separate, 15 pp.

Hofman, H. O., and Demond, C. D. Some Experiments for Determining the Refractoriness of Fire-clays. Trans., Am. Inst. Min. Eng., Vol. XXIV, pp. 42-66, 1894; also Separate, 25 pp.

Holmes, J. A. Notes on the Kaolin and Clay Deposits of North Carolina. Trans., Am. Inst. Min. Eng., Vol. XXV, pp. 929-936. New York, 1896.

Irelan, L. Pottery. 9th An. Rep., California State Mineralogist, pp. 240-261. Sacramento, 1890.

Irving, R. D. The Mineral Resources of Wisconsin. Trans., Am. Inst. Min. Eng., Vol. VIII, pp. 502-506. 1879-'80. (Bricks, Clays and Kaolin.).

Johnston, W. D. Clays. 9th An. Rep., State Mineralogist of California, pp. 287-308. Sacramento, 1890.

Kennedy, W. Texas Clays and Their Origin. Science, Vol. XXII, No. 565, pp. 297-300 Dec. 1, 1893.

Kennicutt, L. P., and Rogers, John F. Fire-clays from Mt. Savage, Allegany County, Maryland. Jour. Analyt. and Applied Chemistry, Vol. V, pp. 101, 542-544, October, 1891; Abstr., Jour. Iron and Steel Inst., Vol. I, p. 306, 1892.

Kerr, W. C. Geology of North Carolina, Vol. I, pp. 296, 297. 1875.

Ladd, G. E. The Clay, Stone, Lime and Sand Industries of St. Louis, Mo., City and County. Bull., Geol. Surv. of Missouri, No. 3, pp. 5-37. St. Louis, 1891.

——————— Notes on the Clays and Building Stones of Certain Western Central Counties, Tributary to Kansas City. Bull., Geol. Surv. of Missouri, No. 5. Jefferson City, 1891.

Langenbeck, K. Chemistry of Pottery, p. 198. Easton, 1896.

Lesley, J. P. Kaolin Deposits of Delaware and Chester Counties, Penn. An. Rep., Pennsylvania Geol. Surv., p. 571. 1885.

Leiber, O. M. Clays. 1st An. Rep., Surv. of South Carolina, pp. 97-99. Columbia, 1858.

McCreath, A. S. Analyses of Clays and Fire-bricks. Rep. MM., 2nd Geol. Surv. of Pennsylvania, pp. 257-279. Harrisburg, 1879.

——————— Analyses of Fire-clays. Rep. M3, 2nd Geol. Surv. of Pennsylvania, pp. 95-97. Harrisburg, 1881.

McGee, W J Brick and Pottery Clays of the Columbia Formation. Report of the Potomac Division, 14th An. Rep., U. S. Geol. Surv., 1892-'93, Part I, pp. 231-232. Washington, 1893.

Mell, P. H.   The Southern Soapstones, Kaolin, Fire-clays and Their Uses. Trans., Am. Inst. Min. Eng., Vol. X, p. 322. (Fire-clay in Alabama). Abstr. in Trans., North of Eng. Inst. Min. Eng., Vol. XXXIII, p. 22. 1883-'84.

Orton, Edward.   The Clays of Ohio, Their Origin, Composition and Varieties. Rep., Geol. Surv. of Ohio, Vol. VII, Part I, Economic Geology, pp. 45-68. Norwalk, 1893.

Orton, Edward, Jr.   The Clays of Ohio and the Industries Established upon Them. Rep., Geol. Surv. of Ohio, Vol. V, pp. 643-721. Columbus, 1884.

———————   The Clay Working Industries of Ohio. Rep., Geol. Surv. of Ohio, Vol. VII, Part I, Economic Geology, pp. 69-254. Norwalk, 1893.

Periodicals.   Thonindustrie Zeitung, Berlin, Germany.

Ries, H.   Clays of Hudson River Valley, 10th An. Rep. of New York State Geologist. 1890. Notes on the Clays of New York State and Their Economic Value. Trans., N. Y. Acad. Sci., Vol. XII, pp. 40-47. December, 1892.

———————   Clay. The Mineral Industry, 1893, Vol. II, pp. 165-210. New York, 1894.

———————   Technology of the Clay Industry. 16th An. Rep., U. S. Geol. Surv., 1894-'95, Part IV, pp. 523-575. Washington, 1895.

———————   Clay Industries of New York. Bull., N. Y. State Mus., Vol. III, No. 12. Albany, March, 1895.

———————   Pottery Industry of the United States. 17th An. Rep., U. S. Geol. Surv., Part III, p. 842.

———————   The Clays of Florida. 17th An. Rep., U. S. Geol. Surv., Part III, p. 871.

———————   Clays of Alabama. Bull., Ala. Geol. Surv. 1897.

———————   The Clay Working-Industry in 1896. 18th An. Rep., U. S. Geol. Surv., Part V, p. 1,105. 1897.

———————   The Ultimate and Rational Analysis of Clays and Their Relative Advantages. Trans., Amer. Inst. Min. Eng., Vol. XXVII.

Russell, I. C.   Sub-aerial Decay of Rocks. Bull., U. S. Geol. Surv., No. 52, pp. 39-43. (Characteristics of Residual Clays.).

Shaler, N. S.   The Brick-making Clays of Massachusetts. 16th An. Rep., U. S. Geol. Surv., 1894-'95. Part II, pp. 324-326. Washington, 1895.

Smith, E. A. The Clays of Alabama. Extr., Proc. Alabama Indus. and Sci. Soc., Vol. II, pp. 33–42, 1892.

Smock, J. C. Mining Clays in New Jersey, Trans., Amer. Inst. Min. Eng., III, p. 211. 1874–'75.

———————— Fire-clays and Associated Plastic Clays, Kaolins, Feldspars and Fire-sands of New Jersey. Trans., Am. Inst. Min. Eng., Vol. VI, p. 177 *et seq.*, 1877–'78. Eng. and Min. Jour., New York, Vol. XXV, pp. 185, 200. 1878.

———————— New Jersey Clays, Ibid., Vol. VI, p. 177. 1879.

Spencer, J. W. Clays. Paleozoic Group of Georgia, Geol. Surv. of Georgia, p. 276. Atlanta, 1893.

Stokes, H. N. Analysis of Kaolin from Georgia Side of Savannah River, near Augusta. Bull., U. S. Geol. Surv., No. 78, p. 120. Washington, 1891.

Struthers, J. Le Chatelier Thermo-electric Pyrometer, School of Mines Quarterly, Vol. XII, p. 143, and Vol. XIII, p. 221.

Tarr, Ralph S. Economic Geology of the United States. New York, 1894. (Clays, pp. 399–402.).

Thompson, Maurice. The Clays of Indiana. 15th An. Rep., State Geologist, pp. 34–40. Indianapolis, 1886.

Tucker, Thomas. American Porcelain. Jour. Franklin Inst., Vol. XXV, p. 43. 1853.

Vulte, H. T. Method of Analyzing Clays. Bull., N. Y. State Mus., Vol. III, No. 12, pp. 141–143. Albany, March, 1895.

Watts, W. L. Placer County (Cali.) 11th Rep., State Mineralogist, pp. 319–322. Sacramento, 1893. (Clay Analyses.).

Wheeler, H. A. The Calculation of the Fusibility of Clays. Eng. and Min. Jour., New York, Vol. LVII, pp. 224–225. March 10, 1894. Abstr., Jour. Iron and Steel Inst., Vol. LXV, pp. 430–432. 1894.

———————— The Fusibility of Clays. Paving and Municipal Engineering, Vol. VI, pp. 200–204. May, 1894.

———————— Vitrified Paving Brick. Indianapolis, 1895.

———————— Clays of Missouri, Missouri Geol. Surv., Vol. XI, pp. 57-67. 1896.

———————— Clays and Shales on the Bevier Sheet. Report on the Bevier Sheet. Missouri Geol. Surv., Vol. IX, pp. 57–67. Jefferson City, 1896.

Willis, Arthur W. Fire-clay. Watt's Dictionary of Chemistry, Vol. II, pp. 651–653. London and New York, 1869.

Williams, Samuel G. Applied Geology. New York, 1886. (Fictile Materials, pp. 319–333.)

Winchell, N. H. Geol. and Nat. Hist. Surv. of Minnesota. Miscellaneous Publications, No. 8, pp. 26–31. St. Paul, 1880. (Cream Colored Brick, p. 28.)

Winslow, Arthur. Report on the Higginsville Sheet. Missouri Geol. Surv., Vol. IX, Jefferson City, 1896. (Clays and Shales, pp. 92–95.)

Worthen, A. H. Economic Geology of Illinois, Vol. I, pp. 278, 453, 470, 496; Vol. II, pp. 55, 303, 316; Vol. III, p. 235.

Young, Jennie J. Ceramic Art. New York, 1878.

INDEX

A

Advisory Board, The, ... 3
Analyses of Georgia Clays,
    108, 115, 126, 133, 135, 139, 146, 185
Anderson, George, The Brick-works of, ... 105
Appendix 193, 199
Augusta Brick Co., The Brick-yard
    of the, ... 152– 153
———— Clays, The, ... 151– 154
————, Potomac Clay near, ... 150
———— Region, The, ... 150
———— Southern R. R., Clays along the, .. 150
————, The Clay Industry of, ... 152– 154

B

Baldwin County 137
Barus, Dr. Carl, Cited 58
Becham, J., The Pottery of, ... 100
————, S., The Pottery of, ... 100
————, W., The Pottery of, ... 100
Behavior of Clays with Reference to Heat,
    Experiments on the, ... 58– 78
————,
    The, ... 84– 85
Behavior of Clays with Reference to Water,
    Summary of Results 56– 57
————,
    Tests of the, ... 44– 57
————,
    The, ... 20– 84
Belair, Clay Deposits along the Georgia
    Railroad East of, ... 162
————, Description of the Crystalline Rocks
    at, ... 159– 160
————, Fire-clay Deposits between Grovetown and, ... 161
Berry's, G. O., Works 94
Bibliography 193– 199
Big Sandy Creek 127
———— District, The, ... 127– 128
Bischof, Cited 74–75, 76
Blake, G. J., The Brick-works of, ... 105
Bond's Store, Potomac Clay near, ... 137
Boswell, Wilbur, The Plant of, ... 158
Burned Clay Products, List of, ... 87
Bussell, J. M., The Pottery of, ... 151

Butler, Clay Openings at, ... 96
———— District, The, ... 96– 97

C

California, General Remarks on the Clays
    of, ... 181– 182
Campfield, H. F., Geological Section on the
    Property of, ... 160
Carr's Station, Potomac Clay at, ... 148, 150
Central of Georgia Railway, Clay Deposits
    along the, ... 103, 105, 111, 128, 131, 137
Chalker, Clay Exposures near, ... 150
————, Exposure of Clay at, ... 151
Chalk Hill, ... 130– 131
Chemical Analysis of Clays, The, ... 78– 79
———— Analyses of Georgia Clays,
    108, 115, 126, 133, 135, 139, 146, 185
———— Analyses, Comparative Table of, . 185
Chemisches Laboratorium für Thonindustrie,
    Clay-testing Furnace Manufactured by
    the, Described 61– 62
Classification of Clays 10– 12
Clays, Geological and Geographical Occurrence of, ... 167– 168
Clay Research by Various States 9
Clays Selected for Study and Comparison .. 169
Coastal Plain, The, ... 85– 91
Colorado, General Remarks on the Clays
    of, ... 180– 181
Columbia Formation, The, ... 88– 91
Columbus 91– 94
————, Clay Industries at, ... 93– 94
Comparison between a Georgia Clay and
    Other Well Known Clays of the United
    States 166– 192
————, Tables of, ... 186– 191
Composition of Clays 15– 18
Cook, Prof. Geo. H., Cited 75, 169
Crawford County Localities, The, ... 97– 100
————, The Small Potteries
    in, ... 99– 100
Cretaceous Period, The, ... 86– 87

D

Davis, John, Potomac Clay on the Farm
    of, ... 129– 130
Definition and Classification of Clays ... 9– 12

(201)

INDEX

Degree of Consolidation of Air-dried Clay, Determination of the.................... 47
Dent, The Small Potteries near,............ 100
Dickson, Thomas, The Pottery of,.......... 100
Dry Branch, Occurrence of Clay at,........ 134

E

Earnest & Co., A. C., The Brick-works of,.. 105

F

Fall Line Belt, General Aspect of the,...89- 91
———— Clays, Eastern Extension of the,. 150
————————, Origin of the,......84- 85
————————, The,................80- 165
Fire-clay, Definition of,..................35, 166
Fitzpatrick, Analysis and Description of Clay from the Vicinity of,.135- 136
————, Clay Exposures in Gullies near,...................134-135, 137
Floyd's Creek, Clays along,................. 150
Fort Valley Plateau, The,................97- 98
Fusibility of Clays, Description of the Electric Furnace Used for Testing,.............66- 67
————, Description of the Oxy-Hydrogen Furnace Used for Testing the,... 67
————, Indirect Method of Testing the,........74- 78
————, New Experiments on the,..............66- 74
————, Use of the Electric Furnace in Testing the, 68- 73
————, Use of the Oxy-Hydrogen Furnace in Testing the..................73- 74

G

Gaillard's, The Small Potteries near,....... 100
General Remarks on Clays 7- 38
Geographical Distribution of Clays 38
Geological Distribution of Clays........... 38
Geology and Physiography of the Fall-Line Clays.....................................81- 91
Georgia, General Remarks on the Clays of,.......................................133- 184
———— Railroad, Clay Deposits along the, 155
————————, Clay Deposits at the 17-mile Post on the,........ 164
————————, Clay Deposits at the 19-mile Post on the,........ 163
————————, Clay Deposits East of Belair along the,............ 162
————————, Clay Deposits near the, . 157

Georgia Railroad, Clay Deposits West of Grovetown along the,... 165
————————, Potomac Clay along the,................148, 150
Glascock County, Potomac Clay in,........ 150
Gordon Road, The Clays along the, ...116- 118
Griswoldville 105
Grovetown District, The,................155- 158
————————, Fire-clay Deposits between Belair and,..... 161

H

Hancock County, Clay Deposits in,........ 148
Harris, Peter, The Brick-works of,......... 105
Hatton, Dr., Cited.......................... 157
Haworth, E., Cited......................... 123
Hofman and Demond, Cited.......58, 76, 77- 78
Huckobee, J. W.,........................... 111

I

Inglett, Mary, The Clay Property of, ...157-158
Irwinton................................... 127

J

James, G. W., Potomac Clay on the Estate of, Described........................ 110
————, Mrs. Sally, Potomac Clays on the estate of, Described.................. 110
Jefferson County, Potomac Clay in,........ 150
Jeffersonville................................ 127
———— Road, Occurrence of Fossiliferous Tertiary Strata Noted 108
————, The White Clay Deposits, near Macon, along the,.............. 108
Johnson and Blake, Cited 30, 31

K

Keg Creek, Clays along,..................... 150

L

Laboratory Investigations... ....40- 79
Lafayette Formation, The,................88- 91
Landis, H. K., Cited 36
Le Chatelier, Cited 15
Letter of Transmittal........................ 5
Lewiston 111
————— Clay Works, The,.........111- 115
—————, Potomac Clay Southeast of,..115- 116
Localities, Notes by,.....................91- 165
Long, H. N., The Pottery of,................ 100

M

Macon & Dublin R. R., Clay Deposits near the, ... 132
Macon Brick Works, The, ... 105
———— District, The, ... 101– 105
————, Potomac Clay at, ... 103
————, The Clay Industry of, ... 104
Magnolia Plains, Clay Deposits at, ... 150
Marshall, H. D., The Pottery of, ... 100
Maryland, The Clays of, ... 171– 172
Massey, Dr. E. I., ... 120
———— Property, Localities South of the, ... 124– 125
———— Property, The, ... 120– 123
Matthews', J. L., Pottery 94
McCants, J. J., The Property of, ... 97
McCoy Brick & Tile Co., The, ... 153– 154
McIntyre Localities, The, ... 128– 180
McMillan, J. W., Clay Works of, ... 148
Merritt, J. N., The Pottery of, ... 100
Methods of Seeking and Testing Clays ... 39– 79
Milledgeville Brick and Pottery Works, The, 148
————, Clay Deposits of, ... 147– 149
————, The Region West of, ... 149
Miller, Z. T., The Clay of, ... 125– 127
Missouri, General Remarks on the Clays of, ... 176– 180

N

Napier's Mill, Potomac Clay in the Vicinity of, ... 136– 137
Newberry, J. S., The Pottery of, ... 100
New Jersey, The Clays of, ... 169– 170

O

Ohio, The Clays of, ... 174– 176
Origin of Clays, ... 12– 14
Orton, Edward, Jr., Cited 174
————, Quoted 175– 176
————, Prof. Edward, Cited 174

P

Park, R. E., Description of the Clay Deposit on the Estate of, ... 102– 103
Payne and Nelson Clay Pit, The, ... 132– 134
Pennington, August, Potomac Clay on the Property of, ... 130
Pennsylvania, General Remarks on the Clays of, ... 172– 173
Piedmont Plateau, The, ... 81– 84
Pine Level, The Small Potteries near, ... 100
Plasticity, Methods of Determining, ... 47– 57
————, Objections to the Indirect Methods of Determining, ... 54– 56
———— of Clays, The, ... 29– 34

Pottery, Typical Small, Described 100
Properties and Characteristics of Clays. 18– 20
Pyrometer, A German Modification of Le Chatelier's, Described 60

R

Raymond's, C. W., Works 94
Redfern, Andrew, The Pottery of, ... 151
————, John, The Pottery of, ... 151
Reed, Silas, Dark-colored Clay on the Property of, ... 156– 157
Rich Hill Localities, The, ... 97– 100
Richmond County, Potomac Clay in, ... 150
Roberts Station, Clay Deposit near, ... 146
Rutley, Description of the Clay Deposit at, . 103

S

Sampling Clays for Laboratory Tests 39– 40
Sandersville, ... 150
Seger and Cramer, Cited 76– 77
———— Cone Method of Measuring High Temperatures Described, The, ... 60– 62
———— Cones in Regulating Temperature in Brick-kilns, The Use of, ... 59, 60
————, Table of Numbers, Composition and Approximate Fusing Temperatures of, ... 63– 65
Shaler, Prof. N. S., ... 43
Shepherd Brothers' Works 94
Shrinkage and Consolidation on Drying .22– 29
———— of Clays on Drying, The, ... 46– 47
Smith & Sons 124
————, Prof. J. Lawrence, Cited 79
————, R. S., the Property of, ... 124
Smock, Prof. John C., Cited 169
Snow, L. A., Potomac Clay on the Property of, ... 129
Southern Railway 132
Special Method for the Separation of Clays 42– 44
Statistics of Valuation of Clays 37
Stevens Pottery 137– 145
Stevens, H., Sons Co.'s Plant, The, ... 104
St. Louis, Mo., The Fire-clays of, ... 178– 180
Stratton, C. C., The Brick-works of, ... 105
Summit, Clay Deposits at, ... 145– 147

T

Ternier, Cited 123
Tertiary Era, The, ... 87– 91
Texas, General Remarks on the Clays of, ... 182– 183
Tool, C. J., The Brick-works of, ... 105
Twiggs County 127
————, Clay Localities in, ... 132– 137

www.ingramcontent.com/pod-product-compliance
Lightning Source LLC
Chambersburg PA
CBHW021939240426
43669CB00047B/549